▪ ORGANIC REACTION MECHANISMS

Second Edition

THE ORGANIC CHEMISTRY

MONOGRAPH SERIES

ORGANIC REACTION MECHANISMS, SECOND EDITION

Ronald Breslow
Columbia University

THE MOLECULES OF NATURE

James Hendrickson
Brandeis University

MODERN SYNTHETIC REACTIONS

Herbert House
Massachusetts Institute of Technology

PEPTIDES AND AMINO ACIDS

Kenneth D. Kopple
Illinois Institute of Technology

INTRODUCTION TO STEREOCHEMISTRY

Kurt Mislow
Princeton University

PRINCIPLES OF MODERN
HETEROCYCLIC CHEMISTRY

Leo A. Paquette
The Ohio State University

STRUCTURE DETERMINATION

Peter Yates
University of Toronto

Second Edition

ORGANIC REACTION MECHANISMS

AN INTRODUCTION

RONALD BRESLOW COLUMBIA UNIVERSITY

W. A. BENJAMIN, INC. ▪ New York

1969

ORGANIC REACTION MECHANISMS: AN INTRODUCTION
Second Edition

Standard Book Numbers: 8053–1252–8 (Cloth)
8053–1253–6 (Paper)

Library of Congress Catalog Card Number 73–80663
Manufactured in the United States of America
12345K32109

The manuscript was put into production on January 13, 1969;
this volume was published on September 5, 1969

W. A. BENJAMIN, INC.. New York, New York 10016

▪ EDITOR'S FOREWORD

UNDERGRADUATE EDUCATION in chemistry is in the midst of a major revolution. Sophisticated material, including extensive treatments of current research problems, is increasingly being introduced into college chemistry courses. In organic chemistry, this trend is apparent in the new "elementary" textbooks. However, it has become clear that a single text, no matter how sophisticated, is not the best medium for presenting glimpses of advanced material in addition to the necessary basic chemistry. A spirit of critical evaluation of the evidence is essential in an advanced presentation, while "basic" material must apparently be presented in a relatively dogmatic fashion.

Accordingly, we have instituted a series of short monographs intended as supplements to a first-year organic text; they may, of course, be used either concurrently or subsequently. It is our hope that teachers of beginning organic chemistry courses will supplement the usual text with one or more of these intermediate level monographs and that they may find use in secondary courses as well. In general, the books are designed to be read independently by the interested student and to lead him into the current research literature. It is hoped that they will serve their intended educational purpose and will help the student to recognize organic chemistry as the vital and exciting field it is.

We welcome any comments or suggestions about the series.

RONALD BRESLOW

New York, New York
December, 1964

▪ PREFACE TO THE FIRST EDITION

IT IS GENERALLY agreed that some study of organic reaction mechanisms should be part of undergraduate organic chemical training. There are several good reasons for this trend: (1) an elementary grasp of how reactions occur is a great help in remembering the factual data which constitute the bulk of an elementary organic chemistry course; (2) an understanding of reaction mechanisms serves as a guide in the design of synthetic sequences; (3) the investigation of organic reaction mechanisms is one of the most active areas of current chemical research, and students should at least be made aware of the existence of this branch of the science; (4) a study of the type of evidence which is used to establish reaction mechanisms introduces the student to a rigorous and stimulating style of thought and induces him to be critical in evaluating the significance of scientific evidence.

Most modern elementary textbooks introduce students to the ideas of reaction mechanisms, and they use mechanistic outlines in describing reactions. However, only in advanced monographs such as Gould's *Mechanism and Structure in Organic Chemistry* or Hine's *Physical Organic Chemistry* are unified outlines of the field of organic reaction mechanisms presented, together with a critical examination of the evidence for proposed mechanisms. In the course of teaching undergraduates the author has long felt the need of a brief introductory book in this area; the present volume is an attempt to fill this need. It is hoped that the book can be read independently by undergraduates who are enrolled in a good course in elementary organic chemistry. It should also be useful in intermediate level courses and for advanced students whose training in this area of organic chemistry was not up to current standards.

In order to solve the problem of providing an introductory treatment and at the same time indicating the current state of research on mechanisms, I have adopted the device of associating a *special topic* with each chapter of the book. While the chapters are general introductions, each special topic deals with an area of current research interest in some detail. General references follow the chapters, but the special topics have footnotes to lead the reader directly to the current research literature. Tables of quantitative data are provided in both chapters and special topics when they seem useful; organic chemistry is still far from being a mathematical science, but the inclusion of a discussion of partial rate factors and the σ^+-ρ relationship in aromatic substitution, for instance, indicates to the student one direction in which the science will certainly grow.

The first chapter on bonding and the second one on methods for determining reaction mechanisms establish the framework for the discussion which follows; each of these chapters is followed by a special topic which further illustrates the chapter material in terms of current research. The remaining five chapters,[1] with their special topics, take up specific classes of reactions and discuss their mechanisms. The criteria used to select these classes of reactions are (1) the reactions are important in synthetic organic chemistry, and (2) a fair amount is known about their mechanisms. Even so, some important reaction types have been neglected (e.g., catalytic hydrogenation, electrophilic aliphatic substitution, photochemical reactions) because of the desire to keep the size and price of this book attractive to its potential readers. However, it is hoped that the choice of topics made will indicate both the scope and depth of current mechanistic theories.

A number of people have read this manuscript at various stages of its preparation, and it is a pleasure to be able to acknowledge

[1] There are six remaining chapters in the second edition.

their helpful comments. Professors Paul Bartlett, Esther Breslow, David Curtin, William S. Johnson, David Lemal, Andrew Streitwieser, and Cheves Walling and Dr. Marjorie Caserio all read the manuscript and made important suggestions at a sacrifice of their own valuable time; my students, Drs. John Brown, Sheila Garratt, Roger Hill, and Edward Robson and Messrs. Lawrence Altman, David Chipman, and Edmond Gabbay not only read and commented on the manuscript, but also tolerated the diversion of my attention away from their research problems. Finally, I wish to thank Mrs. Helga Testa for her outstanding services as typist.

RONALD BRESLOW

New York, New York
April 1965

▪ PREFACE TO THE SECOND EDITION

Although less than four years has elapsed since publication of the first edition of this book, progress in the field of reaction mechanisms has made a second edition essential. In this new edition we have brought the previously covered topics up to date and added new material on ribonuclease, the Brønsted relationships, hydroboration, and other areas.

The most striking advance since the first edition appeared is the emergence of the ideas of Woodward and Hoffmann on orbital symmetry and its role in determining reaction rates and stereochemistry. These ideas are now briefly introduced in Special Topic 1 and 4 and form the subject of a new Special Topic 8. A new Chapter 8 on Photochemistry has also been added in recognition of the developing mechanistic patterns in this branch of chemistry.

Helpful comments from a number of users of the book have guided the revisions. I would particularly like to acknowledge the contributions of Prof. Thomas Spencer of Dartmouth College and Prof. Robert Bergman of California Institute of Technology for critical reviews of the first edition, and Prof. Nicholas Turro of Columbia University for his review of the new Chapter and Special Topic 8.

RONALD BRESLOW

New York, New York
June 1969

CONTENTS

1

▪ BONDING IN ORGANIC COMPOUNDS

THE PRIMARY CONCERN of this book is with chemical reactions and their mechanisms. However, much of the evidence on reaction mechanisms is derived by considering the relative reactivities of similar compounds; changes in a molecule which speed up or slow down one of its reactions tell us much about the mechanism of that reaction. In order to understand the relationship between structure and reactivity it is first necessary to review certain aspects of valence theory.

1–1 Simple Molecules

Methane. Bonding between two atoms can be described by saying that putting an electron pair in the space between two positive nuclei "cements" them together. In these terms methane can be symbolized as is shown below, the line simply representing an electron-pair bond.

```
     H
     |
  H—C—H
     |
     H
```

On a higher level of sophistication the bonds are described in terms of atomic orbitals and their overlap. Thus carbon has two electrons in its inner 1*s* shell which are not involved in bonding, and its outer valence shell can accommodate up to eight electrons in the 2*s* and three 2*p* orbitals, two electrons in each. These are shown below (the *x*, *y*, and *z* axes are arbitrary in direction but must be mutually perpendicular).

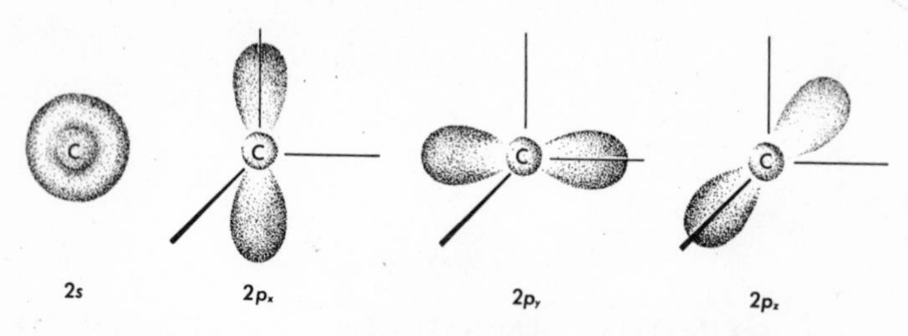

Carbon atomic orbitals.

The 2*s* orbital has two parts, an inner sphere and an outer shell, and the *p* orbitals also have two parts, or *lobes,* projecting on opposite sides of the carbon nucleus. In methane these four atomic orbitals are hybridized into four tetrahedral orbitals, which are called sp^3 orbitals. These have somewhat the form of a *p* orbital, but the mathematical parts of valence theory indicate that adding the 25% of *s* character should build up one lobe of a *p* orbital and diminish the other; the quantitative treatment furthermore indicates that these four orbitals will be tetrahedrally oriented. Bonding occurs by overlap of each of these tetrahedral orbitals with a hydrogen 1*s* orbital.

In methane the four orbitals are completely equivalent, but in less symmetrical molecules hybridization may occur so as to put more *s* character in some orbitals and more *p* character in others.

An sp^3 orbital. Tetrahedral set of four sp^3 orbitals.

An important consideration is that a carbon 2*s* orbital is 5.3 electron volts (122 kcal/mole) lower in energy than is a 2*p* orbital; accordingly, hybrid orbitals with more *s* character will attract electrons more strongly. For instance, in methyl chloride the very electronegative chlorine atom attracts the electrons of the C—Cl bond. The methyl group can release these electrons somewhat by using a hybrid orbital for bonding with less than 25% *s* character, so that the orbital is less electron-attracting, but then the extra *s* character shows up in the three C—H bonds.

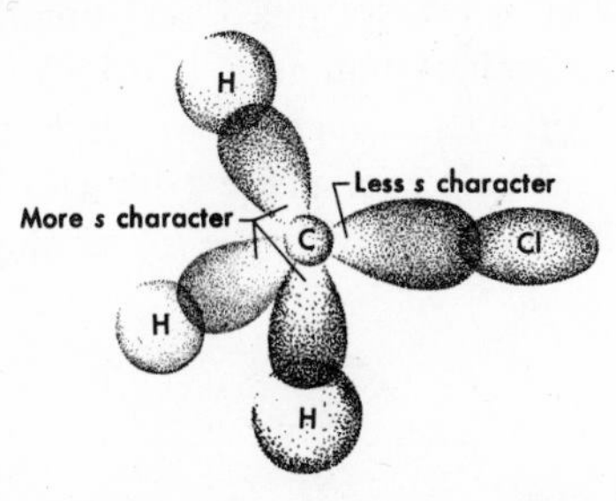

This results in a decreased electron density at the protons compared with that in methane, since the more electro-negative carbon orbital (with extra *s* character) pulls the bonding electrons closer.

Geometrical factors can also cause a departure from perfect tetrahedral hybridization. For example, in cyclopropane the ring angles of 60° are much smaller than the 109° between tetrahedral orbitals; on the other hand, simple *p* orbitals make an angle of 90°, which is much closer to the ring angle. Accordingly, in cyclopro-

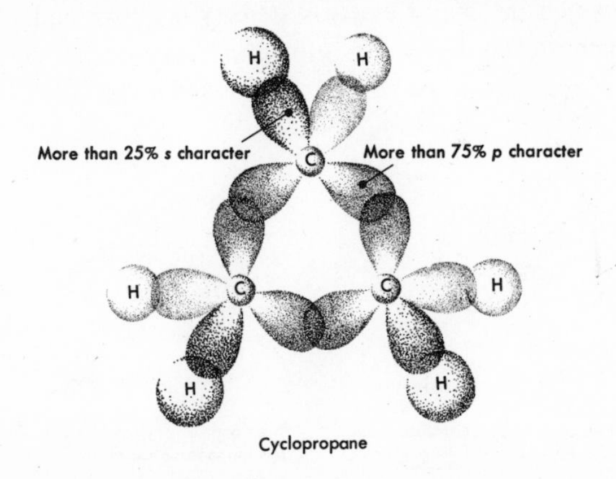

Cyclopropane

pane the ring bonds are made using hybrid orbitals with more p character; the bonds to hydrogen thus have more than 25% s character, so they are shorter than the C—H bonds in methane.

It is apparent that very few saturated molecules will have perfect tetrahedral hybridization, but for most purposes this is ignored and bonding by saturated carbon is stated to involve the use of sp^3 orbitals.

Ethylene. One could imagine a description of the ethylene double bond in terms of overlap of two tetrahedral orbitals from each carbon. In contrast to this "bent-bond" approach is a description in which each carbon atom hybridizes the s orbital with only two of the p orbitals. Such an sp^2 hybridized carbon will have three hybrid orbitals in a plane (the xy plane if p_x and p_y were used) at 120° angles, with the remaining p orbital (p_z)

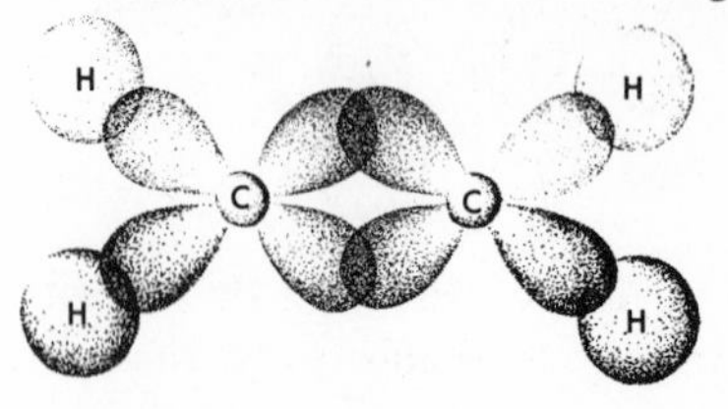

The bent-bond description of ethylene.

perpendicular to the plane. Then the double bond between carbons will consist of one ordinary single bond from overlap of an sp^2 orbital from each carbon (a σ bond, or straight-line bond, in which the major electron density is along the internuclear line) and one special bond formed by sideways overlap of the p_z orbitals (a π bond, in which the major electron density is above and below the internuclear line).

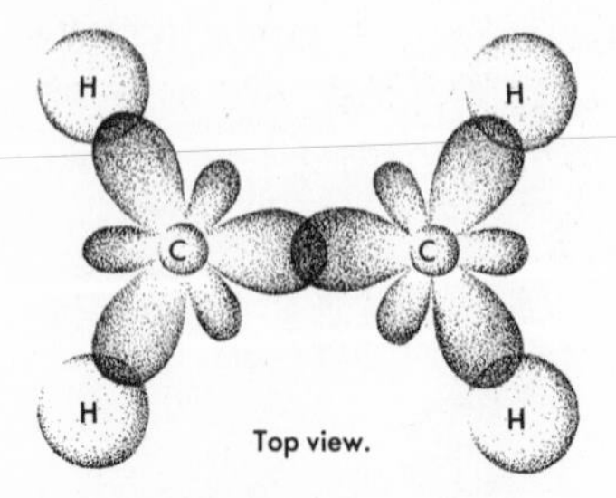

The σ bonds of ethylene. The molecule lies in the plane of the paper and the p_z orbitals are perpendicular to it.

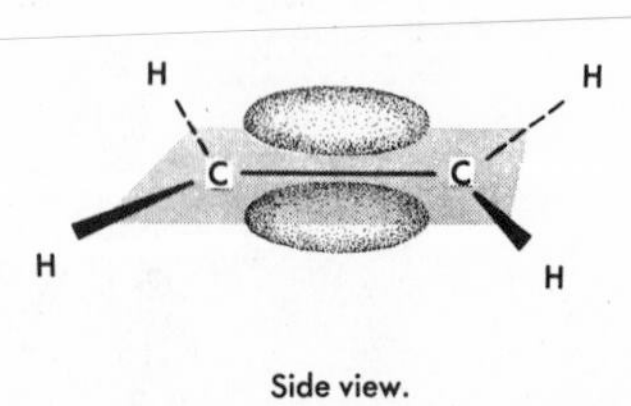

The π bond of ethylene. This bond is produced by the combination of the p_z orbitals of each carbon.

One difference between these two modes of description is that the bent-bond model predicts an H—C—H angle of 109°28′ while the σ—π double-bond model predicts 120°. The actual value is 117°, intermediate but close to the predictions of the usual σ—π description.

The *p* orbital can be exactly described by a mathematical function, and the square of the function gives the electron densities which we have been picturing. The function itself has a positive sign in one lobe of the *p* orbital and a negative sign in the other lobe (and a value of zero at the nucleus). Two parallel *p* orbitals may thus be arranged in two possible orientations, in one of which lobes of the same mathematical sign are together, while in the other the functions are of opposite sign.

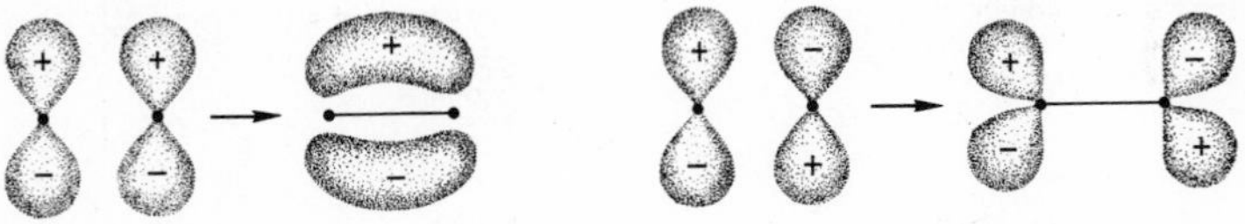

$\frac{1}{\sqrt{2}}(p_1 + p_2)$
bonding

$\frac{1}{\sqrt{2}}(p_1 - p_2)$
antibonding

π orbitals of ethylene

The first combination (formed by adding the mathematical functions for the two *p* orbitals and multiplying the sum by $1/\sqrt{2}$, a normalizing factor) forms the stable *bonding* π orbital into which the two electrons of the π bond are placed. The second combination is *antibonding* and leads to a high-energy molecular orbital; this is vacant in normal ethylene but would be occupied in an electronically excited state, as is discussed in Chapter 8.

Ethylene has no dipole moment. Even though the C—H bond is undoubtedly somewhat polar the geometry of the molecule is such that the four C—H dipoles cancel; the same geometrical principle that C—H dipoles will cancel can be shown for saturated hydrocarbons, even quite complex ones. However, propylene has a dipole moment of 0.34 Debye units, and this reflects a slight drift of electrons from the methyl group toward the unsaturated carbon.

0.34 Debye units.

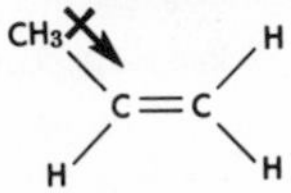

CH_3, H, C=C, H, H

This suggests that one carbon atom is more electronegative than another one, which is at first surprising. However, the important point is the electronegativity of the orbitals, since the electrons will drift away from the sp^3 orbital of saturated carbon and into the sp^2 orbital of the unsaturated carbon. A $2s$ orbital is of lower energy than a $2p$ orbital; consequently electrons will be more stable in the orbital with more s character.

Acetylene. The C—C triple bond could again be described in terms of bent bonds, but the more usual description involves a σ bond and two π bonds. Thus the carbon $2s$ orbital is hybridized

The two *sp* hybrid orbitals at an acetylenic carbon, each with one large and one small lobe.

The σ bonds of acetylene.

The π bonds of acetylene. The molecule is really cylindrically symmetrical.

with one of the $2p$ orbitals; the result is two sp hybrid orbitals pointing in opposite directions along the same line. These two are used to construct the σ bonds to hydrogen and carbon, while the remaining two $2p$ orbitals on each carbon are used for π bonds.

1–2 Conjugated Molecules

Benzene. The idea of the electron pair bond between adjacent atoms accounts poorly for the structure of benzene. Here the carbons are again sp^2 hybridized, all carbons and hydrogens lying in a plane (the xy plane). The remaining p orbitals (p_z orbitals) are all parallel to each other and perpendicular to the molecular plane. One could perhaps imagine preferential π bonding of a p orbital with only one of its neighbors if the bond distances were not equal, but since all carbon-carbon distances are 1.39 Å, orbitals should pair equally well on either side. This is indicated in valence bond theory by retaining the idea of the electron pair bond but writing benzene as a resonance hybrid of two "canonical" structures. These latter simply indicate pairing schemes for the orbitals, and the resonance arrow indicates that the real structure involves both ways of pairing.

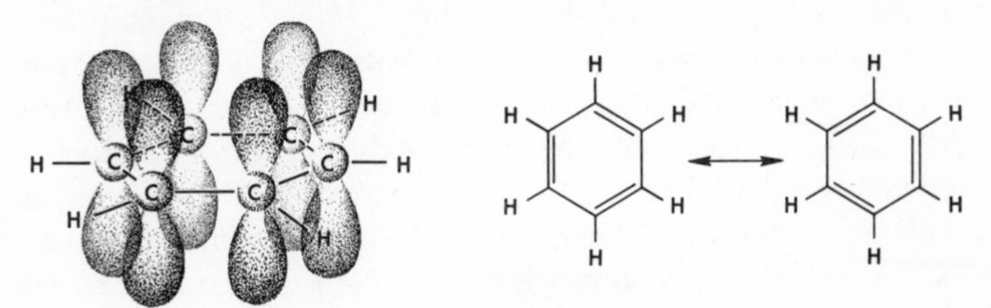

The atomic p_z orbitals in benzene.

A resonance hybrid of two different pairing schemes.

Some other pairing schemes can be imagined, involving overlap across the ring. These will contribute less to the structure of benzene, although they are taken into account in quantitative valence bond calculations.

A set of higher energy canonical structures (resonance forms) which contribute only slightly to the structure of benzene.

Such resonance structures imply no change of geometry; they simply represent ways of pairing orbitals in the molecule which has the flat symmetrical structure of benzene.

The fact that benzene does not have three isolated double bonds, but instead a delocalized π electron system, means that it is considerably more stable than might have been otherwise expected. The amount of this stability is indicated by heats of hydrogenation. When an ordinary double bond is hydrogenated 28.6 kcal/mole of heat is evolved under certain standard conditions. If benzene had three ordinary double bonds it should thus evolve 85.8 kcal/mole on reduction to cyclohexane; in fact only 49.8 kcal/mole are given off. This means that benzene contains 36 kcal/mole less energy than might have been expected if it were not conjugated (this 36 kcal/mole is called its "resonance energy"). Valence bond theory accounts for this extra stability.

Resonance interaction between canonical (hypothetical) structures is stabilizing. The more resonance forms (stable pairing schemes) can be written for any molecule the more stabilized the actual compound will be; unstable resonance forms contribute little to either the structure or the stability. If we count only those structures without long bonds or ionic bonds (the stable structures are often called "Kekulé forms"), we can account for the resonance energies (amounts of extra stability) of a number of aromatic systems (Table 1–1). Here we have drawn only one Kekulé form for each compound. This is a common practice which we shall follow throughout the book to save space; the actual structure should of course be symbolized by drawing all the important resonance forms, interconnected by resonance arrows.

TABLE 1–1 ▪

Resonance Energies versus *Number of Resonance Forms*

Molecule	Number of Kekulé Forms	Resonance Energy in Kcal/Mole
 Benzene	2	36
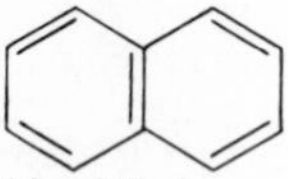 Naphthalene	3	61
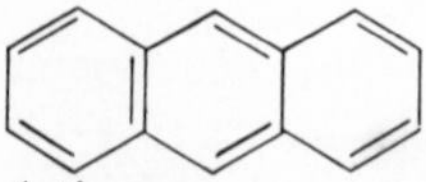 Anthracene	4	83.5
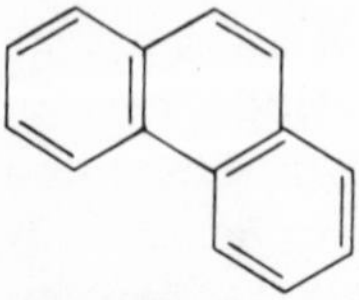 Phenanthrene	5	91.3

There is an alternative way of describing molecules, by the *molecular orbital* approach to molecular quantum mechanics. This is quite different from valence bond theory in that it abandons entirely the idea of an electron-pair bond. Although simple molecules such as methane can also be described in molecular orbital terms mathematically, there is little useful difference in the *qualitative* picture which emerges. However, for conjugated systems molecular orbital theory has some striking advantages. We shall apply it in a simple and semiempirical fashion; it should be emphasized that intense research in theoretical chemistry today is concerned with serious calculations of molecular properties by the application of molecular orbital methods.

Let us assume that the single bonds in benzene, carbon-to-carbon as well as carbon-to-hydrogen, are adequately described already in terms of the electron-pair bond. However, instead of placing the six π electrons in atomic orbitals and only then considering pairing and exchange interaction, let us first combine the p_z atomic orbitals into molecular orbitals, encompassing all the carbon atoms. Then we place the π electrons directly in these new orbitals. Three such orbitals will be needed to accommodate the

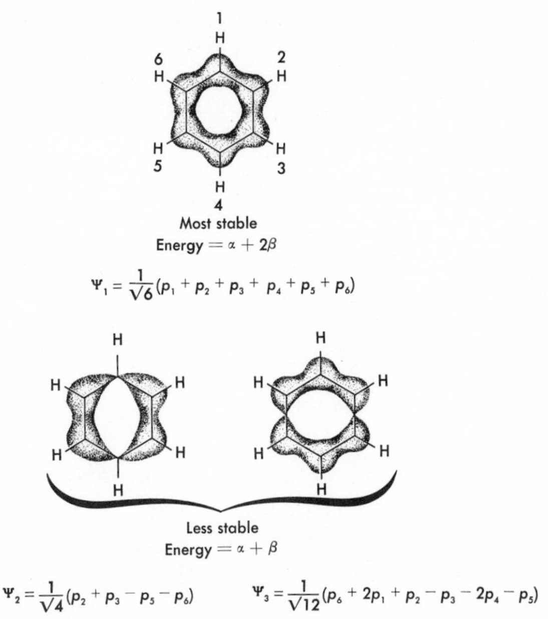

Occupied benzene π molecular orbitals.

six π electrons, two in each, and by quantitative application of the theory (and unfortunately not otherwise) the shape and energy of these three orbitals can be derived.

There are also three unstable molecular orbitals which ordinarily play no role. However, when benzene absorbs ultraviolet light an electron is excited from a stable orbital into an unstable one (cf. Chapter 8). With quantitative molecular orbital calculations it is possible to predict the energy required for this process, and thus to predict the ultraviolet spectrum of the compound.

The energies of the three occupied benzene orbitals are given in terms of two quantities, α and β: α, the *coulomb integral,* is the energy of an ordinary isolated carbon $2p$ orbital, while β, the *resonance integral,* is a negative energy whose value cannot be conveniently computed; β is therefore evaluated from experimental data such as heats of combustion or hydrogenation. In the lowest energy benzene molecular orbital, $E = \alpha + 2\beta$, an electron is more stable by 2β (a negative energy), because it is delocalized over all six atoms, than it would be if it were isolated in a single carbon $2p$ orbital (with energy $= \alpha$).

This orbital, Ψ_1, has all six p_z orbitals lined up with positive lobes above and negative lobes below the plane of the ring, so each p–p interaction is bonding. In orbital Ψ_3 the p_z orbitals on carbons 6, 1, and 2 are lined up with positive lobes up, but 3, 4, and 5 have positive lobes down. The result is that the region of the molecule

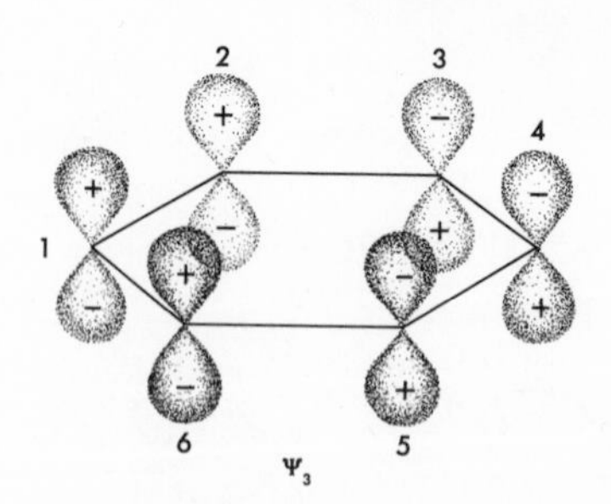

Ψ_3

between atoms 6, 1, and 2 (and also between atoms 3, 4, and 5) is a bonding region, where electrons are stable, but between atoms 2 and 3 (and 5 and 6) there is an antibonding region. This is like the antibond in the excited-state orbital of ethylene, and corresponds to a region in which electrons are of higher energy. In fact, precisely halfway between carbons 2 and 3 there is a node where the π molecular orbital has a value of zero (in passing from a positive function at carbon 2 to a negative function at carbon 3), and at the node the electron density is zero. The mathematical form of Ψ_3 corresponds to this picture: p_6, p_1, and p_2 have positive signs, while p_3, p_4, and p_5 have negative signs. Electron density tends to be greater in the center of each bonding region, at carbons 1 and 4, as is shown by the fact that p_1 and p_4 have twice the weight of the other individual orbitals in the expression for Ψ_3. The $1/\sqrt{6}$, $1/\sqrt{4}$, and $1/\sqrt{12}$ are normalizing factors which indicate that Ψ_1, Ψ_2, and Ψ_3 are weighted averages of the p_z orbitals rather than simple sums.

The orbital Ψ_2 is similar in having two nodes, but these come at carbons 1 and 4 rather than between adjacent atoms. Since the function has a value of zero at these two atoms, p_1 and p_4 do not appear in the expression for Ψ_2. Benzene has six-fold symmetry, and it may at first seem strange that carbons 1 and 4 have been singled out as different from the others. However, the extra electron density at carbons 1 and 4 in Ψ_3 makes up for the zero density in Ψ_2, and with six electrons placed two each in Ψ_1, Ψ_2, and Ψ_3 the total electron density is six-fold symmetric. The orbitals Ψ_2 and Ψ_3 have the same energy, $\alpha + \beta$, and both are less stable than Ψ_1, since nodes and antibonding regions raise the energy of an orbital.

With two electrons in each of these three orbitals, the total π electron energy is $6\alpha + 8\beta$.

An ordinary double bond also has its electrons delocalized, over two carbons; the energy of each such electron in an ordinary double bond is computed to be $\alpha + \beta$. Thus, if benzene had three ordinary double bonds the total π electron energy would be $6\alpha + 6\beta$. A real benzene molecule is therefore predicted to be 2β more stable ($[6\alpha + 8\beta] - [6\alpha + 6\beta]$) than the hypothetical unconjugated cyclohexatriene. The 2β of stabilization is often equated with the -36 kcal/mole derived from heats of hydrogenation, so that β is then set equal to -18 kcal/mole. Although a number of gross oversimplifications are involved in this theory, it is reasonably successful at correlating the resonance energies of various aromatic systems (Table 1–2).

Butadiene. In this open-chain molecule conjugation effects play only a small role. Thus, while it used to be thought that the short C_2—C_3 bond (1.483 Å compared with 1.526 Å for the

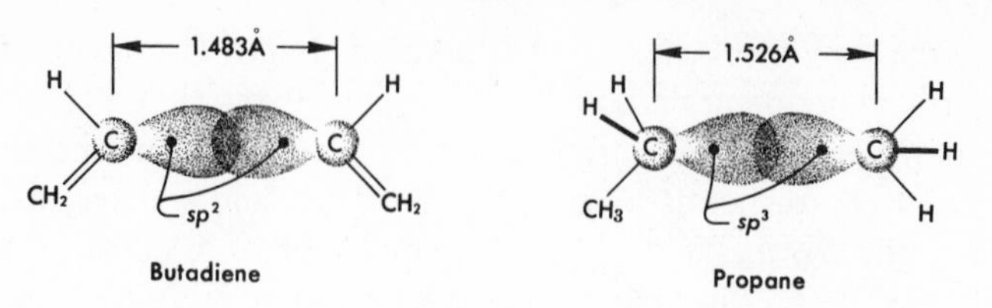

TABLE 1–2 ▪

Resonance Energies Calculated by Simple Molecular Orbital Theory

Compound	Calculated Resonance Energy β's	Kcal/Mole ($\beta = 18$ Kcal/Mole)	Measured Resonance Energy in Kcal/Mole
Benzene	2.00	36.0	36.0
Naphthalene	3.68	61.4	61.0
Anthracene	5.31	95.6	83.5
Phenanthrene	5.45	98.1	91.3
Triphenylene	7.28	131.0	117.7

C—C single bonds of propane) was due to double bond character, it is now clear that most of the shortening is due to hybridization. The C_2—C_3 bond is essentially a single bond, but a single bond involving (more or less) sp^2 hybrid orbitals; with more *s* character in the orbitals such a bond is shorter than that made by saturated carbons.

Measured heats of hydrogenation indicate only 3.5 kcal/mole of resonance energy for butadiene, and some of this apparent "resonance energy" is probably also ascribable to other effects. Accordingly, only one pairing scheme has major significance in butadiene; resonance forms with long bonds do not contribute much to the structure.

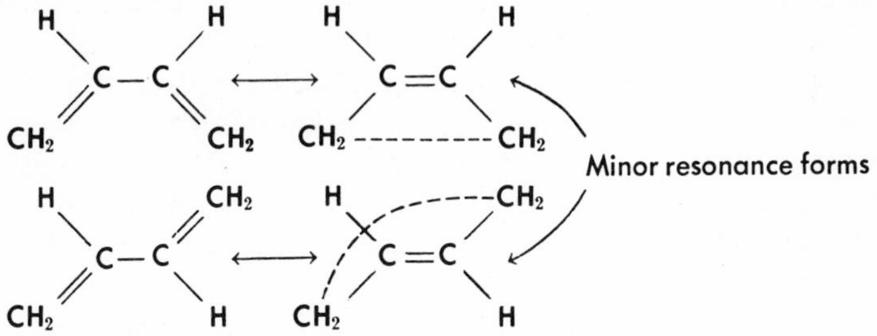

Minor resonance forms

When a proton is added to butadiene, however, the situation changes. The resulting carbonium ion has two almost equivalent resonance forms, and its structure is a hybrid of the two.

$$CH_2{=}CH{-}CH{=}CH_2 + H^{\oplus} \longrightarrow CH_3{-}\overset{\oplus}{C}H{-}CH{=}CH_2 \longleftrightarrow CH_3{-}CH{=}CH{-}\overset{\oplus}{C}H_2$$

Many other reactions of butadiene also proceed through intermediates which are strongly resonance-stabilized. Thus, even though butadiene itself does not show major conjugation effects, such effects may become important in its reactions. In the above case, for instance, the formation of a strongly stabilized (by conjugation) carbonium ion from a weakly stabilized diene makes the reaction favorable. Such effects are most strikingly seen when one considers the relative strengths of organic acids and bases.

1–3 Organic Acids and Bases

In water solution a proton acid ionizes by proton transfer to the solvent. An equilibrium constant can be measured for this process, and the negative logarithm of this equilibrium constant is called the pK_a of the acid. It is convenient to think of the pK_a as the pH of a solution in which the acid is 50% ionized or neutralized; thus weak acids will have a high pK_a, since they will require a high pH (strong alkali) to lose their protons.

$$HA + H_2O \leftrightarrows H_3O^{\oplus} + A^{\ominus}$$

$$pK_a = \text{pH when } \frac{[A^-]}{[HA]} = 1$$

It is not possible to measure pK_a's in water if they are much outside the range of 0 to 14, since water itself is both an acid and a base. An acid so strong that its pK_a is below 0 will simply transfer all its protons to water, forming the new acid H_3O^+, while if the acid is so weak that its pK_a lies above 14 a base will not remove its proton, but instead will attack water to form OH^-. However, it is possible to extrapolate outside this region using other data. In Table 1–3 are listed pK_a's for some compounds of particular interest.

Since the ionization of a strong acid yields a weak base (called the conjugate base of the acid which formed it), and vice versa, Table 1–3 can also be interpreted in terms of base strengths. Thus it shows us that the methyl anion is an extremely strong base while the nitrate ion is not very basic, and it also indicates that aniline is a weaker base than is methylamine.

The position of an equilibrium depends on the relative stabilities of the species on both sides of the equation, and the exact nature of this dependence can be described in mathematical form:

$$\Delta G^\circ = -2.3RT \log K$$

Here ΔG° is the energy change, called the standard free energy change, on going from starting materials to products; K is the equilibrium constant of the reaction and T is the absolute temperature, while R is the gas constant. For our purposes it will be enough to note that if one substitutes the correct numbers this

TABLE 1–3 ▪

Approximate pK_a's of Some Proton Acids

Compound		pK_a
	C—H IONIZATIONS	
$CH_3—H$	Methane	58
$C_6H_5CH_2—H$	Toluene	37
$C_6H_5C\equiv C—H$	Phenylacetylene	21
$CH_3C(=O)CH_2—H$	Acetone	20
$H—CH_2NO_2$	Nitromethane	10.2
$H—CN$	Hydrogen cyanide	9.14
$(CH_3C(=O))_2CH—H$	Acetylacetone	9.0
$H—CH(NO_2)_2$	Dinitromethane	3.6
	N—H IONIZATIONS	
$NH_2—H$	Ammonia	36
$C_6H_5NH—H$	Aniline	27
$CH_3\overset{\oplus}{N}H_2—H$	Methylammonium ion	10.64
Pyridinium ($C_5H_5\overset{\oplus}{N}H$)	Pyridinium ion	5.17
$C_6H_5\overset{\oplus}{N}H_2—H$	Anilinium ion	4.58
	O—H IONIZATIONS	
$(CH_3)_3CO—H$	*t*- Butyl alcohol	20
$CH_3O—H$	Methanol	16
H_2O	Water	15.7*
$C_6H_5O—H$	Phenol	9.95
$CH_3CO_2—H$	Acetic acid	4.76
$HCO_2—H$	Formic acid	3.77
$CF_3CO_2—H$	Trifluoroacetic acid	– 0.25
$H—ONO_2$	Nitric acid	– 1.37

* Corrected from 14 since H_2O concentration is 55 moles/liter.

equation indicates that, at room temperature, a lowering of one unit in the pK_a of an acid results if its ionization is made 1.4 kcal/mole more favorable. In interpreting the pK_a's of acids, then, we consider the relative stabilities of the acid and of the conjugate base.

Toluene is much more acidic than methane, and this is almost entirely due to conjugation in the benzyl anion.

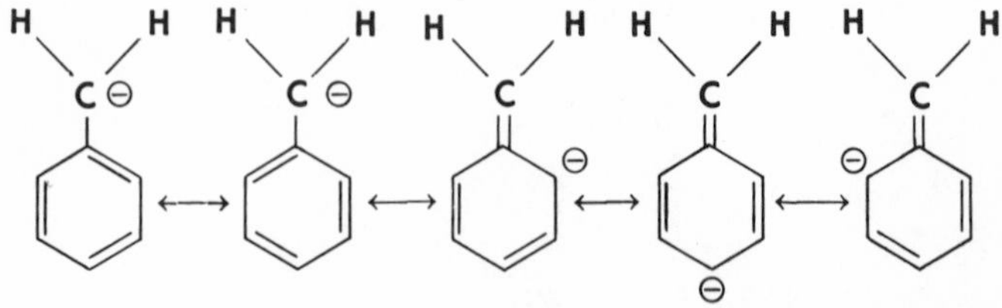

Of course the first two resonance forms shown are ordinary benzene resonance forms, and such resonance would stabilize toluene as well. The last three forms, however, play a role only in the anion, so it is they alone which stabilize the anion *relative* to toluene, and thus contribute to the extra acidity. If all the extra acidity of toluene compared with methane is due to this resonance interaction, and if the pK_a's are correct, the 21 units difference corresponds to 29 kcal/mole (1.4×21) of extra resonance stabilization in the benzyl anion. This effect of a phenyl group is also seen in the acidity of aniline compared with ammonia, and in the pK_a's of phenol and of the anilinium ion, but the effect is much smaller. In the phenoxide ion, for instance, the negative charge is more likely to stay on the electronegative oxygen; consequently the resonance forms in which the charge is found in the benzene ring become less important.

Acetone and nitromethane are also acidified because their anions are conjugated systems whose true structures can be represented as hybrids.

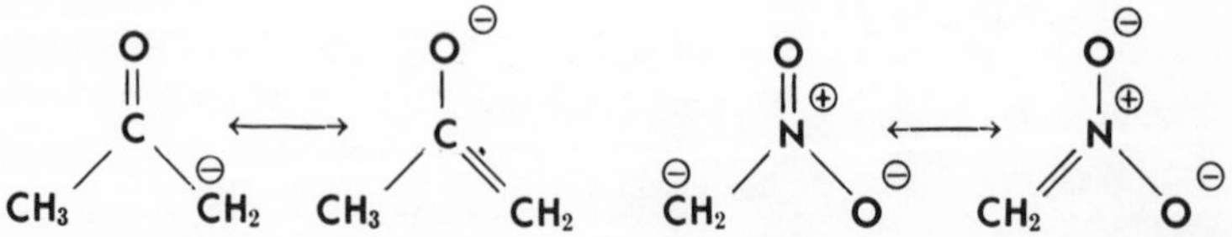

Of course in these cases it is not only this resonance interaction which matters, but also the fact that in the major resonance form the negative charge is on oxygen rather than on carbon. Acetylacetone and dinitromethane illustrate the strong effects observed when two such activating groups are present. A resonance explanation can also be offered for the acidity of acetic acid. Even though acetic acid itself has the same *number* of resonance forms as does the acetate ion, resonance in the ion is more stabilizing since the two forms are of equivalent energy, which is not the case for acetic acid. Resonance forms which involve the

$$CH_3-C(=O)-O^{\ominus} \longleftrightarrow CH_3-C(-O^{\ominus})=O$$

$$CH_3-C(=O)-O-H \longleftrightarrow CH_3-C(-O^{\ominus})=\overset{\oplus}{O}-H$$

development of charge separation generally do not contribute as much.

Other factors also play a role in determining acidity. Solvation effects can be quite important, as several examples in Table 1–3 illustrate. Although methanol and water are of comparable acidity, *t*-butyl alcohol is a weaker acid. This probably reflects the fact that it is difficult to solvate the anion of *t*-butyl alcohol since the oxygen is so closely surrounded by methyl groups; water can surround the charged oxygens in methoxide or hydroxide ions more easily, and stabilize them by hydrogen bonding and orientation of solvent dipoles. The fact that formic acid is more acidic than is acetic acid has also been shown to be due mostly to better solvation of the formate ion; one side of the carboxylate group is shielded in acetate ion by the methyl.

Occasionally a substituent can *assist* in solvation and thus make a compound *more* acidic. Salicyclic acid has a pK_a of 2.98 while benzoic acid has a pK_a of 4.20. Comparison with other compounds shows that most of the acidification by the *o*-hydroxyl

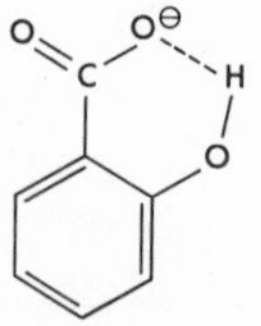

Salicylate ion.

group is the result of internal hydrogen bonding of the carboxylate ion, the OH group thus acting like a solvent molecule.

Polar groups can favor anion formation by *inductive electron-withdrawing effects*. Thus trifluoroacetic acid is more acidic than is acetic acid, and this is partly due to electron shifts in σ bonds; there is a general drift of electrons toward the fluorine atoms, making the oxygen atoms slightly electron-deficient and thus better able to accommodate a negative charge. In addition to this factor there is a very important interaction which has been called a *field effect*. A negative charge can be solvated by bringing some positively charged particle into the vicinity, or by orienting a solvent dipole so that its positive end is nearby. The field effect is an intramolecular example of the same thing; the carbon-fluorine dipole is oriented so that its positive end is near the carboxylate ion, and this dipole-charge interaction helps stabilize the ion. Such field effects of polar groups are quite common; another example may be found in the nitrophenols. Since the pK_a of *p*-nitrophenol is 7.14 and that of *m*-nitrophenol is 8.35, both are more acidic than phenol itself (9.95). In the *p*-nitrophenoxide ion a resonance form is important in which the negative charge is placed in the nitro group, but in *m*-nitrophenol this is not possible and the major stabilization of the anion comes from interaction of the negative charge with the dipole. Such an interaction also plays some role in *p*-nitrophenoxide ion.

Inductive effects and field effects also play a role in some of the systems which we have discussed only in terms of resonance; thus obvious polar effects are expected in the carbonyl and nitro compounds in our list. Part of the lowered basicity of aniline compared with methylamine (increased acidity of the anilinium ion compared with the methylammonium ion) is due to the fact that phenyl groups have an electron-withdrawing inductive effect compared with methyls. This is due to orbital electronegativity

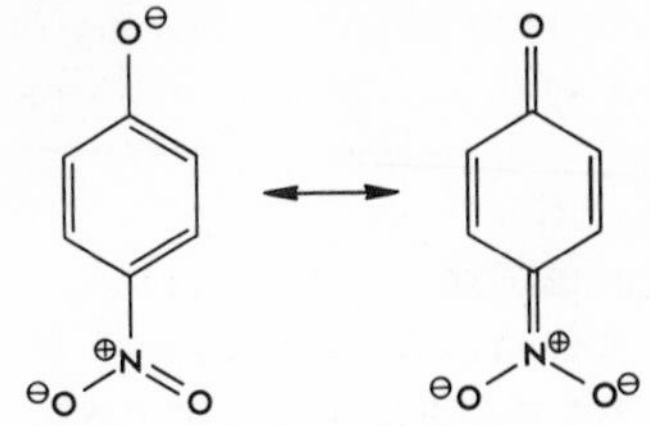

Conjugative stabilization of *p*-nitrophenoxide ion.

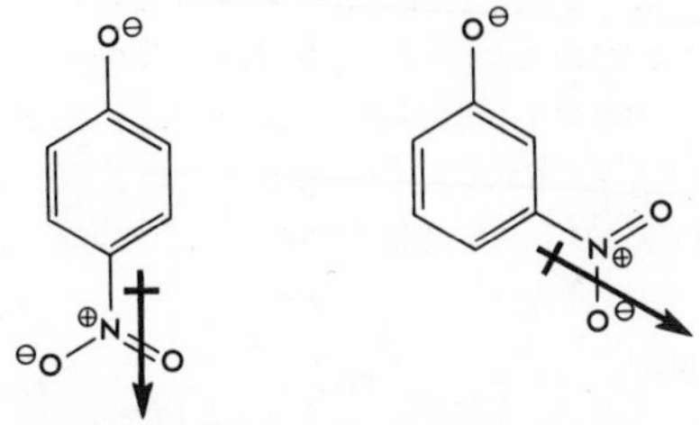

Ion-dipole interaction in *p*- and *m*-nitrophenoxide ions.

(an sp^2 orbital from phenyl compared with an sp^3 orbital from methyl in the σ bond). Orbital electronegativity can also show up more directly in determining acidity. Acetylenes are acidic because the C—H σ bond involves an *sp* orbital from carbon; this *sp* orbital accommodates an electron pair, and thus a negative charge, quite well because of the lower energy of the *s* orbital. A similar effect shows up in HCN, along with an inductive effect of nitrogen. Hydrogens on double bonds are also acidified by orbital electronegativity, and the fact that pyridine is a weaker base than methylamine is largely due to the presence of the unshared electron pair of pyridine in an sp^2 orbital.

Although by implication the factors which affect base strength have been included in our discussion of acids, a few special points should be mentioned. Amides are much less basic than amines. Protonation on the nitrogen of an amide would interfere with the amide resonance.

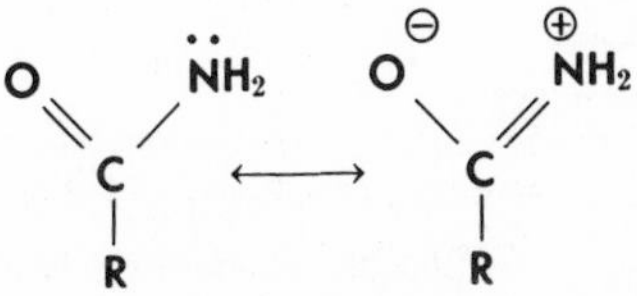

Amide resonance

However, because of this resonance an amide actually protonates on oxygen in strong acid; the protonated amide has even more resonance stabilization, since now there are *two* resonance forms with no charge separation.

It should be apparent that simple resonance arguments are not adequate to let us predict whether *this* process, which does not interfere with the amide resonance, will be easier or harder than the protonation of a simple amine, since the two reactions are not really related. The same situation is met in guanidine. Protonation on an amino group would interfere with the guanidine resonance, but protonation actually occurs on the imino nitrogen, leading to a strongly stabilized cation. Because of this, guanidine turns out to be a very strong base.

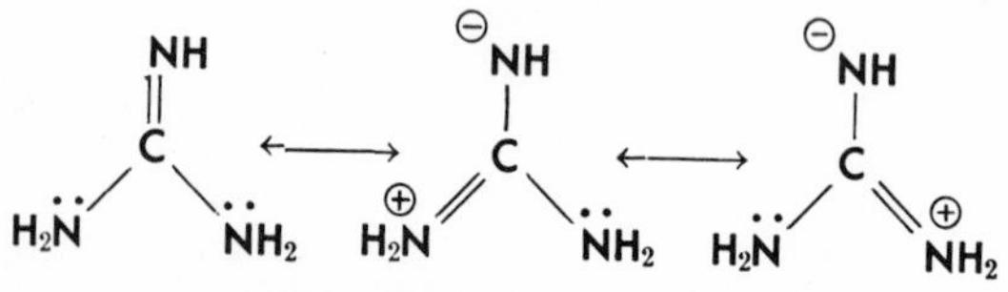

Resonance in guanidine

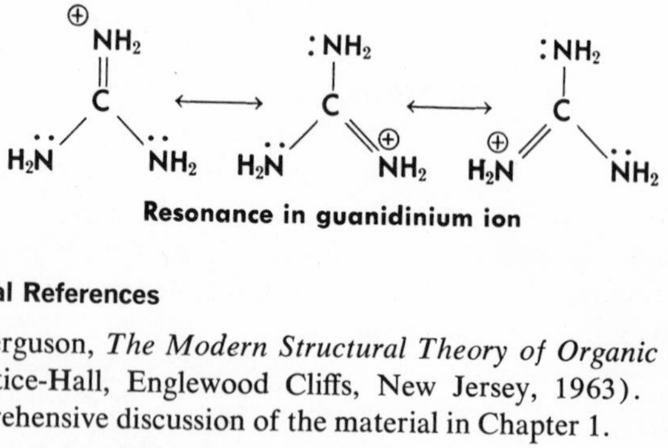

Resonance in guanidinium ion

General References

L. N. Ferguson, *The Modern Structural Theory of Organic Chemistry* (Prentice-Hall, Englewood Cliffs, New Jersey, 1963). The most comprehensive discussion of the material in Chapter 1.

L. Pauling, *The Nature of the Chemical Bond* (3rd ed., Cornell University Press, Ithaca, New York, 1960). A good general introduction.

G. W. Wheland, *Resonance in Organic Chemistry* (John Wiley & Sons, New York, 1955). A general book on bonding in organic compounds. Particularly useful for material on resonance energies and on acid-base strengths. The last chapter is an introduction to the mathematics of valence bond and molecular orbital calculations.

C. A. Coulson, *Valence* (Oxford University Press, London, 1961). A rather mathematical treatment, but many sections will be comprehensible and interesting even to the student with little mathematical training.

H. B. Gray, *Electrons and Chemical Bonding* (W. A. Benjamin, New York, 1964). A good general introduction, chiefly concerned with inorganic compounds.

J. D. Roberts, *Notes on Molecular Orbital Calculations* (W. A. Benjamin, New York, 1961). The best simple introduction to m.o. calculations for organic chemists.

A. Streitwieser, Jr., *Molecular Orbital Theory for Organic Chemists* (John Wiley & Sons, New York, 1961). An extensive treatment. A fair amount of background is needed to understand some sections, but the descriptive material which outlines the results of m.o. calculations can be understood by any chemist.

H. A. Bent, "An Appraisal of Valence-Bond Structures and Hybridization in Compounds of the First-Row Elements," *Chemical Reviews,* **61,** 275 (1961). A discussion of hybridization, with emphasis on the ways in which bond distances and angles depend on the exact hybridization involved.

M. Dewar, *Hyperconjugation* (Ronald Press, New York, 1962). A critical attack on the conclusions of simple molecular orbital theory, in which it is argued that many of the effects which have been ascribed to "conjugation" are due instead to the dependence of single-bond properties on hybridization.

I

Special Topic

AROMATICITY AND THE $4n + 2$ RULE[1]

BENZENE IS CONSIDERED to be an "aromatic" compound because its cyclic conjugation leads to great stability (for reasons we have discussed in both resonance and m.o. terms). "Aromaticity" in this sense of the word thus means special stability due to cyclic conjugation. Other definitions of aromaticity are sometimes used which focus attention on some of the other special properties of benzene and its relatives, but we shall restrict ourselves to considering it in terms of conjugative stabilization (not total stability, but *that part of it* which is due to cyclic conjugation).

The various substituted derivatives of benzene, such as phe-

1. For good reviews on this topic, cf. (a) A. Streitwieser, Jr., *Molecular Orbital Theory for Organic Chemists* (John Wiley & Sons, New York, 1961), Chapter 10; (b) M. F. Vol'pin, "Non-benzenoid Aromatic Compounds and the Concept of Aromaticity," *Russian Chemical Reviews,* 1960, p. 129; (c) D. Lloyd, *Carbocyclic Non-Benzenoid Aromatic Compounds* (Elsevier Publishing Co., New York, 1966).

nol, are aromatic compounds as well since they contain the benzene system, but other aromatic systems are also known. Naphthalene, anthracene, and phenanthrene derivatives contain new aromatic systems, but these are still related to benzene since they simply contain several benzene rings fused together. Other aromatic compounds contain nitrogen analogs of benzene, such as pyridine or pyrimidine rings, either singly or fused to other aromatic rings, but again these represent only a slight extension of the benzene concept.

A slightly different situation is found in the aromatic furan, pyrrole, and thiophene rings, and such extensions as the oxazole, imidazole, and thiazole systems.

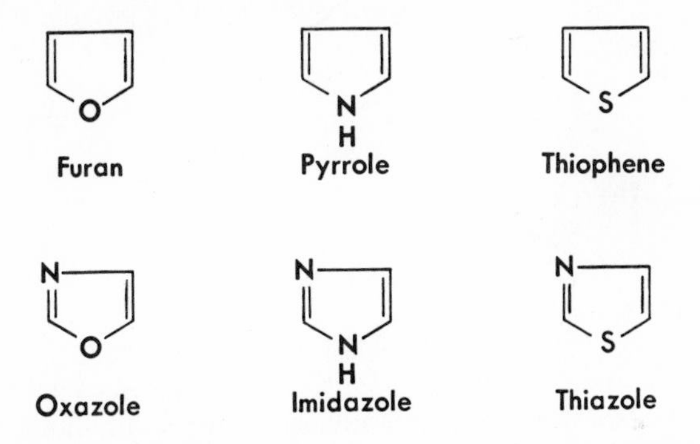

The resonance energies (conjugative stabilizations) found[2] from heats of combustion are furan, 16 kcal/mole; pyrrole, 21 kcal/mole; and thiophene, 28 kcal/mole. Thus all three systems are strongly stabilized by conjugation, although not by as much as is benzene. In resonance terms this stabilization is due to the contribution of a series of charge-separated forms.

Resonance forms for the five-membered heterocycles.

The fact that pyrrole is more aromatic than is furan can be ascribed to a greater resonance contribution when the positive

2. G. Wheland, *Resonance in Organic Chemistry* (John Wiley & Sons, New York, 1955), p. 98.

charge is on nitrogen than when it is on the more electronegative oxygen. The even greater stabilization in thiophene is usually[3] attributed to the fact that unlike carbon, oxygen, nitrogen, etc., sulfur is not totally restricted to having eight electrons in its valence shell, since more electrons may be placed in the vacant 3*d* orbitals. An extra resonance form for thiophene in which the sulfur has expanded its valence shell to ten electrons is shown below.

Closely related to these heterocycles is cyclopentadienyl anion, another aromatic system.

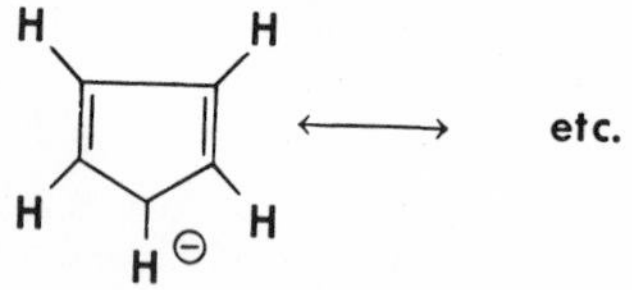

It is not really convenient to determine the resonance energy of this anion by hydrogenation or combustion, but a good indication is found in the fact[4] that cyclopentadiene has a pK_a of about 20. At 1.4 kcal/mole for each pK unit, the difference in acidity between cyclopentadiene and methane corresponds to a large resonance stabilization of the cyclopentadienyl anion.

Another fundamental aromatic system is the cycloheptatrienyl cation, often called the "tropylium" ion.[5]

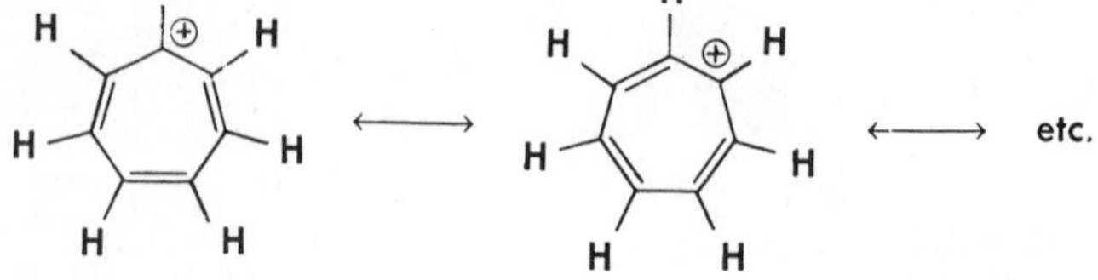

3. G. Cilento, "The Expansion of the Sulfur Outer Shell," *Chemical Reviews,* **60,** 147 (1960).
4. Ref. 1a, p. 414.
5. W. E. Doering and L. H. Knox, "The Cycloheptatrienylium (Tropylium) Ion," *Journal of the American Chemical Society,* **76,** 3203 (1954).

Here too the most convenient indication of aromaticity is a pK. A carbonium ion may react with water in a reversible equilibrium to form the corresponding alcohol.

$$R^+ + 2H_2O \rightleftarrows ROH + H_3O^+$$

$$pK_{R^+} = pH \text{ when } \frac{[R^+]}{[ROH]} = 1$$

Again a pK may be defined for this reaction as the negative log of the equilibrium constant. This is commonly[6] called a pK_{R^+} to indicate that it involves a carbonium ion, but again it is simply equal to the pH when the carbonium ion is half neutralized, so it is very similar to a pK_a. The pK_{R^+} of tropylium ion is + 4.7,[5] showing that this carbonium ion is so stable that even in water solution it only becomes converted to the alcohol when the pH is raised above 4.7, i.e., when the solution approaches neutrality. An unconjugated carbonium ion would be very reactive toward water. Even triphenylmethyl cation, in which the positive charge is conjugated with three phenyl rings, has[7] a pK_{R^+} of − 6.63, so it is much less stable than tropylium ion and exists only in *very* strong acid solutions.

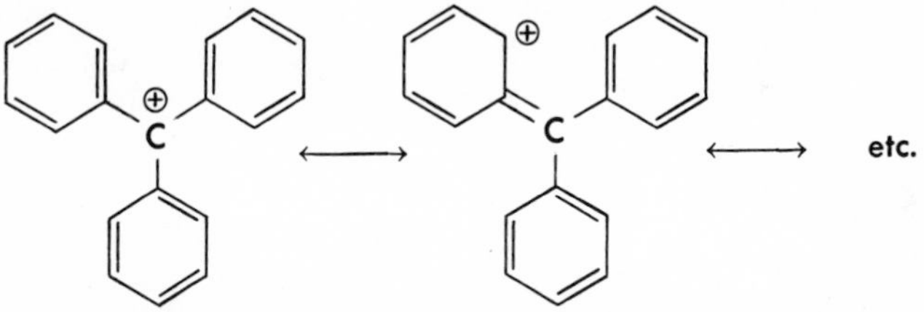

⊕ C ⟷ ⊕ C ⟷ etc.

Triphenylmethyl cation, $pK_{R^+} = -6.6$

It might seem from these examples that any cyclic system will be aromatic if the electrons are delocalized so that a number of resonance forms contribute to the structure. Interestingly, this is not the case. The cycloheptatrienyl anion is extremely unstable and has only recently been prepared.[8] The same instability is apparent in the cyclopentadienyl cation;[1] it has not even been

6. Ref. 1a, p. 363.
7. N. Deno, J. Jaruzelski, and A. Schriesheim, "Carbonium Ions. I.," *Journal of the American Chemical Society*, **77**, 3044 (1955).
8. H. Dauben and M. Rifi, "Cycloheptatrienide (Tropenide) Anion," *Journal of the American Chemical Society*, **85**, 3041 (1963).

possible to prepare this, although a number of derivatives are known. The pK_{R^+} of pentaphenylcyclopentadienyl cation[9] is approximately -16, indicating instability so extreme that the cation can exist only in acids stronger than concentrated H_2SO_4!

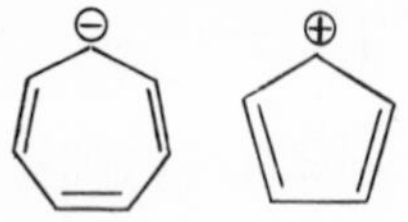

Nonaromatic systems

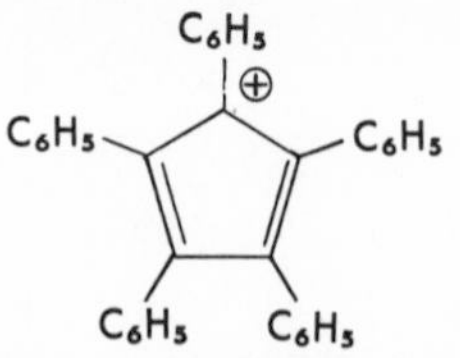

Pentaphenylcyclopentadienyl cation, $pK_{R^+} -16$

What can possibly explain this situation? A five-membered anion or a seven-membered cation is aromatic, but when the charges are reversed very unstable systems are formed. An early suggestion[10] called attention to the fact that all of the aromatic systems contain six π electrons (considering the fused ring aromatic compounds as simple extensions of the monocyclic systems). Thus the idea was put forward that an "aromatic sextet" of π electrons had special properties. This idea would also explain why cycloöctatetraene is not aromatic. Even though the compound is not planar[11] it seemed surprising that it did not prefer to flatten in order to gain aromatic stability; but it contains eight π

Cycloöctatetraene, nonaromatic and nonplanar.

9. R. Breslow and H. Chang, "The Rearrangement of Pentaphenylcyclopentadienyl Cation," *Journal of the American Chemical Society,* **83,** 3727 (1961); R. Breslow, H. W. Chang, R. Hill, and E. Wasserman, "Stable Triplet States of Some Cyclopentadienyl Cations," *Journal of the American Chemical Society,* **89,** 2135 (1967).
10. J. Armit and R. Robinson, "Polynuclear Heterocyclic Aromatic Types," *Journal of the Chemical Society,* **127,** 1604 (1925).
11. W. Person, G. Pimentel, and K. Pitzer, "The Structure of Cycloöctatetraene," *Journal of the American Chemical Society,* **74,** 3437 (1952).

electrons, so according to the sextet postulate it would not be expected to be aromatic even if it *were* planar.

This idea also accounts for the overwhelming evidence now available[12] that cyclobutadiene is a very unstable substance, which can have a transient existence at best. On inspection it would seem that this compound has two Kekulé structures and that it should be stabilized by being a hybrid of the two, just as benzene was.

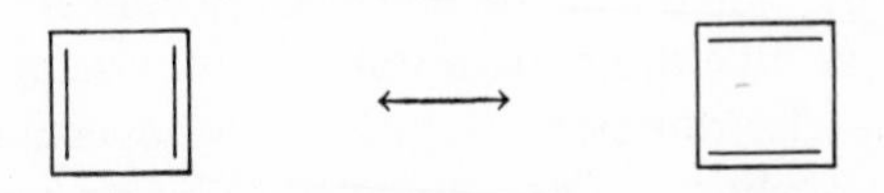

Resonance forms for cyclobutadiene, a nonaromatic system

However, it has four π electrons instead of the magic sextet.

The requirement of a sextet of electrons for aromaticity cannot be explained by valence bond theory. However, if we examine the way in which cyclobutadiene is described in molecular orbital terms it becomes possible to understand why the compound is so different from benzene in its properties. The four π electrons of cyclobutadiene are placed in the m.o.'s shown.[13]

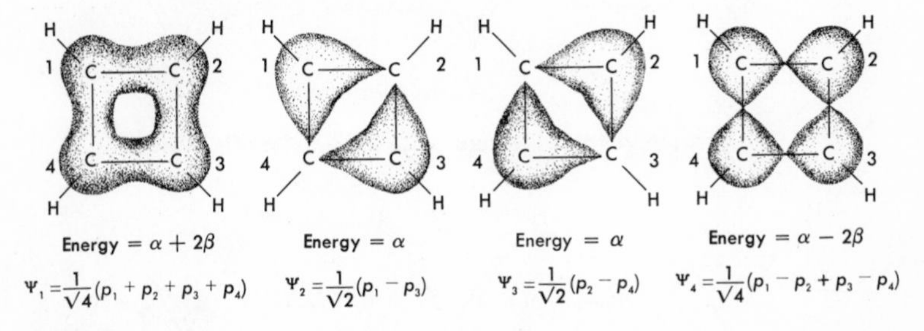

All four m.o.'s are pictured, although since four π electrons are present the orbital of highest energy will not be occupied. Two electrons go into Ψ_1, and the other two go, one each, into Ψ_2 and Ψ_3. Thus, the total π electron energy for four electrons is $4\alpha + 4\beta$; for two isolated double bonds it would also be $4\alpha + 4\beta$, so cyclobutadiene has no resonance energy. This certainly explains

12. M. P. Cava and M. J. Mitchell, *Cyclobutadiene and Related Compounds* (Academic Press, New York, 1967).
13. Ref. la, p. 62.

why it is not aromatic, although other factors must be responsible for the observed high instability.

Unfortunately, this result depends in entirety on the fact that quantitative m.o. calculations do predict the energies indicated for these orbitals, and a fair background in quantum mechanics is required to make these calculations comprehensible. When such calculations are done on a variety of conjugated systems an interesting pattern emerges. The monocyclic four π electron systems, such as cyclobutadiene and cyclopentadienyl cation, are predicted to have little or no extra stability because of conjugation, while the six π electron systems such as cyclopentadienyl anion, benzene, and tropylium ion are predicted to be strongly stabilized. Thus the magic effect associated with a sextet emerges as a prediction of m.o. theory. Strikingly, this is not the only magic number for which aromaticity is predicted. The calculations predict that the cyclopropenyl cation, a two π electron cyclic conjugated system, should also be aromatic while its four π electron relative, the cyclopropenyl anion, should not.[14]

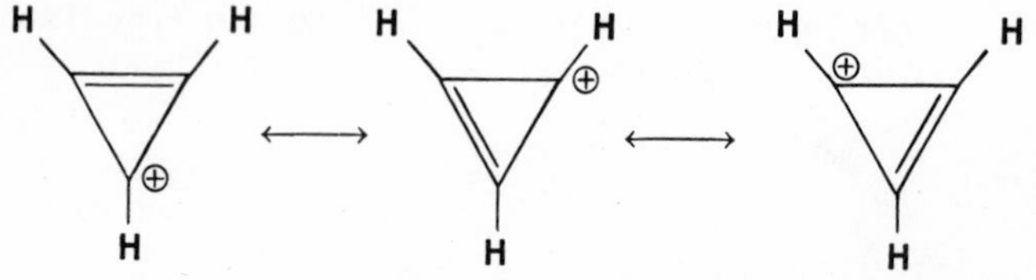

Cyclopropenyl cation, a two π electron aromatic system

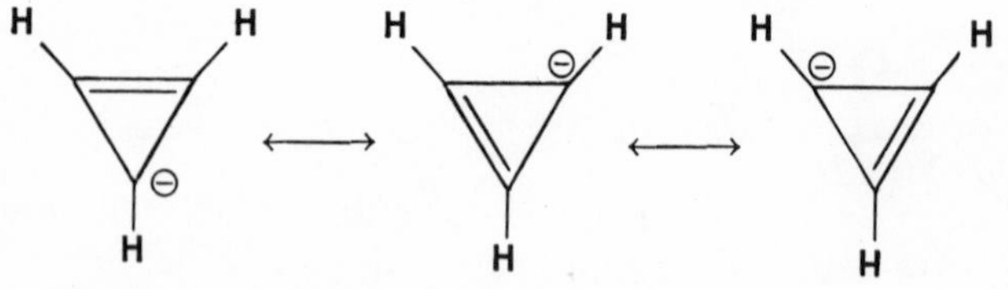

Cyclopropenyl anion, a four π electron *nonaromatic* system

This result can also be understood easily once we see the stabilities predicted for the cyclopropenyl m.o.'s. Three such

14. J. Roberts, A. Streitwieser, and C. Regan, "Molecular Orbital Calculations of Properties of Some Small-Ring Hydrocarbons and Free Radicals," *Journal of the American Chemical Society,* **74,** 4579 (1952).

m.o.'s are formed by combining the three atomic p_z orbitals of the ring carbons.

In the cyclopropenyl cation, two electrons are placed in the

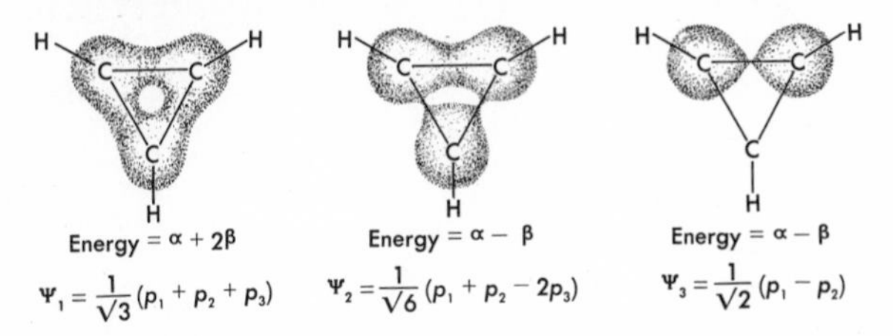

H
C
Energy = α + 2β
$\Psi_1 = \frac{1}{\sqrt{3}}(p_1 + p_2 + p_3)$
Energy = α − β
$\Psi_2 = \frac{1}{\sqrt{6}}(p_1 + p_2 - 2p_3)$
Energy = α − β
$\Psi_3 = \frac{1}{\sqrt{2}}(p_1 - p_2)$

The three m.o.'s of the cyclopropenyl system

lowest m.o., so the total π electron energy is $2\alpha + 4\beta$. If the system were unconjugated its π energy would be $2\alpha + 2\beta$ (two electrons in an ordinary double bond) so resonance stabilizes it by 2β, the same as the resonance energy of benzene. In the cyclopropenyl anion, two more electrons will be added to the next two m.o.'s; because they happen to have the same energy each of these m.o.'s will accept one electron and will be half-filled as in cyclobutadiene. The total π electron energy for the anion is thus $4\alpha + 2\beta$. If the system were unconjugated it would have a π electron energy of $4\alpha + 2\beta$ ($2\alpha + 2\beta$ for the double bond, 1α each for the two electrons in an isolated p orbital); thus, the cyclopropenyl anion has no extra resonance energy, and it is not aromatic.

These predictions have been confirmed. A number of very stable derivatives of the cyclopropenyl cation are known, and the parent compound has recently been prepared.[15] Thus triphenylcyclopropenyl cation[16] has a pK_{R^+} of + 3.1 while tripropylcyclopropenyl cation[17] has a pK_{R^+} of + 7.2, and is thus stable even in

15. R. Breslow, J. T. Groves, and G. Ryan, "Cyclopropenyl Cation," *Journal of the American Chemical Society*, **89,** 5048 (1967); D. G. Farnum, G. Mehta, and R. G. Silberman, "Ester Decarbonylation as a Route to Cyclopropenium Ion and Its Mono- and Dimethyl Derivatives," *Journal of the American Chemical Society*, **89,** 5048 (1967).
16. R. Breslow and H. Chang, "Triarylcyclopropenium Ions," *Journal of the American Chemical Society*, **83,** 2367 (1961).
17. R. Breslow, H. Hover, and H. Chang, "The Synthesis and Stability of Some Cyclopropenyl Cations with Alkyl Substituents," *Journal of the American Chemical Society*, **84,** 3168 (1962).

neutral water. This is the most stable hydrocarbon cation known.

In further agreement with these predictions, it is found that the cyclopropenyl anion system is not aromatic; it is so unstable that no derivatives have yet been isolated, although they have been

C_6H_5, C_6H_5, C_6H_5 — C=C, C^+ ⟷ etc.

Triphenylcyclopropenyl cation,
$pK_{R^+} = +3.1$

$CH_3CH_2CH_2$, $CH_2CH_2CH_3$, $CH_2CH_2CH_3$ — C=C, C^+ ⟷ etc.

Tripropylcyclopropenyl cation,
$pK_{R^+} = +7.2$

detected[18] as transient intermediates. In fact, the experimental data strongly indicate that cyclopropenyl anion is *antiaromatic,* i.e., destabilized by conjugation.[18]

The situation in the cyclopropenyl system can be contrasted with the calculated energies of the corresponding open-chain analogs, allyl cation and allyl anion. Three molecular π orbitals can also be constructed in this system: a stable bonding orbital Ψ_1, an unstable antibonding orbital Ψ_3, and a nonbonding orbital Ψ_2 whose energy is the same as that of a simple p_z orbital.

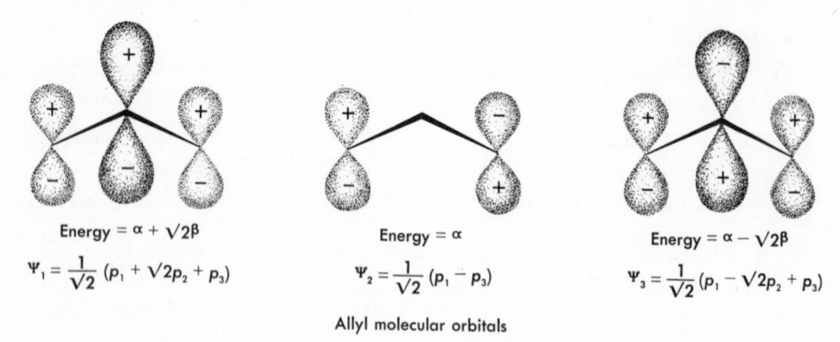

Energy = $\alpha + \sqrt{2}\beta$
$\Psi_1 = \frac{1}{\sqrt{2}}(p_1 + \sqrt{2}p_2 + p_3)$
Energy = α
$\Psi_2 = \frac{1}{\sqrt{2}}(p_1 - p_3)$
Energy = $\alpha - \sqrt{2}\beta$
$\Psi_3 = \frac{1}{\sqrt{2}}(p_1 - \sqrt{2}p_2 + p_3)$

Allyl molecular orbitals

The allyl cation has two electrons in Ψ_1, and a total π electron energy of $2\alpha + 2\sqrt{2}\beta$. The cation is thus less stabilized than cyclopropenyl cation, with π energy $2\alpha + 4\beta$. In the allyl anion

18. R. Breslow, J. Brown, and J. Gajewski, "Antiaromaticity of Cyclopropenyl Anions," *Journal of the American Chemical Society,* **89,** 4383 (1967).

two more electrons are placed in Ψ_2, and the total π energy is $4\alpha + 2\sqrt{2}\beta$. This anion is more stabilized than is the cyclopropenyl anion, which has π energy $4\alpha + 2\beta$.

This relationship between the cyclic and open-chain cations and anions can also be seen by considering *orbital symmetries.* The allyl cation has its two electrons in Ψ_1, a molecular orbital with all three p_z orbitals aligned in the positive direction. Thus in the (conceptual) conversion of the allyl to the cyclopropenyl cation, by joining carbon 1 to 3, the interaction between p_1 and p_3 will be bonding, so the cyclic system will be more stabilized. On the other hand, in Ψ_2 orbitals p_1 and p_3 have opposite symmetry: p_1 has its positive lobe up while p_3 has its positive lobe down. When allyl anion is conceptually converted to cyclopropenyl anion, the two electrons in Ψ_1 develop a bonding interaction as above, but the two electrons in Ψ_2 develop an antibonding interaction. Since the coefficients of p_1 and p_3 are larger in Ψ_2 $(1/\sqrt{2})$ than in Ψ_1 $(1/2)$, the antibonding effect dominates, and cyclopropenyl anion is less stable than allyl anion. Of course, this symmetry discussion has given us only a qualitative idea of the situation, compared with the direct quantitative use of the molecular orbital energies. However, we will see that this kind of qualitative treatment can have wide utility.

When m.o. calculations of this type are done for a variety of possible aromatic compounds, including systems not yet known, it is found that not only 2 and 6, but also 10, 14, 18, etc. electrons may form monocyclic aromatic systems. These numbers fit the form $4n + 2$, where $n = 0, 1, 2, \ldots$, and this generalization has been called the "$4n + 2$ rule" for aromaticity, or Hückel's rule after its proponent.[19]

Considering the case where $n = 2$, i.e., ten π electron systems, it seems that one obvious possibility would be cyclodecapentaene. In a ten-membered ring with five *cis* double bonds, delocalized to form an aromatic system, the internal angles would be 144° compared with the sp^2 optimal angle of 120°; this involves some strain, although probably not a prohibitive amount. If two *trans* double bonds are used the angle strain disappears, but at the expense of severe hindrance between the two hydrogen atoms in the middle of the ring. Neither system has yet (1968) been

19. E. Hückel, "Quantentheoretische Beitrage zum Benzolproblem," *Zeitschrift für Physik,* **70,** 204 (1931).

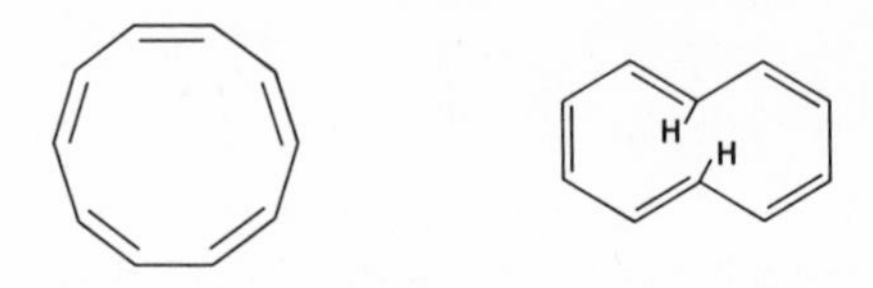

Two possible cyclodecapentaenes.

prepared.[20] On the other hand, two simple ten π electron compounds are now known, both of which are apparently aromatic. One is the cycloöctatetraenyl dianion;[21] only the aromaticity of the system explains why it is relatively easy to place two negative charges in the same ring.

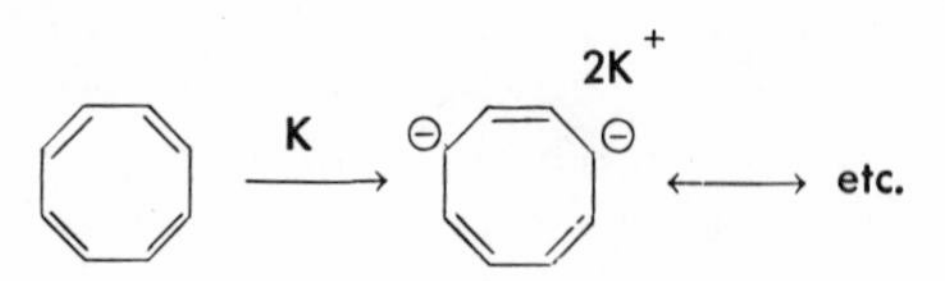

Cyclooctatetraenyl dianion, a 10 π electron aromatic system

The other system is the cyclononatetraenyl anion,[22] again a stable delocalized system.

Cyclononatetraenyl anion

Vogel has prepared some ten π electron systems with bridging groups.[23] While these are not monocyclic, the π electrons are

20. For detection of an isomer, see E. van Tamelen and T. Burkoth, "Cyclodecapentaene," *Journal of the American Chemical Society,* **89,** 151 (1967).
21. T. Katz, "The Cycloöctatetraenyl Dianion," *Journal of the American Chemical Society,* **82,** 3784 (1960).
22. T. Katz and P. Garratt, "Cyclononatetraenyl Anion," *Journal of the American Chemical Society,* **85,** 2852 (1963); E. La Lancette and R. Benson, "Cyclononatetraenide: An Aromatic 10-π-Electron System," *Journal of the American Chemical Society,* **85,** 2853 (1963).
23. E. Vogel *et al., Angewandte Chemie, International Edition,* **3,** 228, 642, 643 (1964).

restricted to the periphery of the system, and the compounds are aromatic by spectroscopic and chemical criteria.

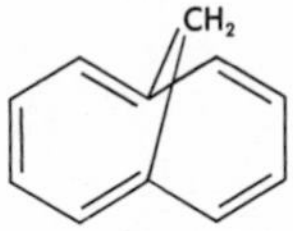

It should be noted that naphthalene has 10 π electrons, as does azulene. These aromatic compounds are not directly relevant to the $4n + 2$ rule, which is rigorously derived only for monocyclic systems. However, the same type of molecular orbital calculation which led to the $4n + 2$ rule can be applied to individual polycyclic systems, and their aromaticity can also be accounted for in molecular orbital terms.[1]

One example of a 14 π electron system is the unusual compound

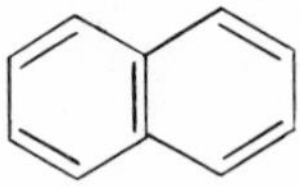

Naphthalene 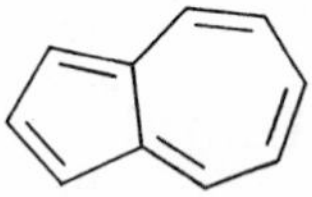 **Azulene**

I.[24] This is a polycyclic compound, but Hückel's rule still applies since the π electron system is monocyclic, i.e., is confined to the 14-membered ring around the edge of the molecule.

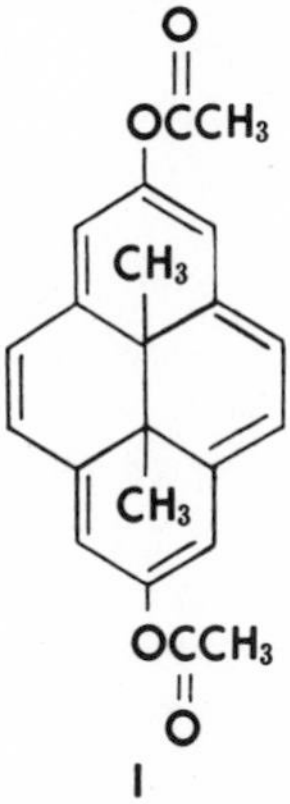

I

24. V. Boekelheide and J. Phillips, "2,7-Diacetoxy-*trans*-15,16-

Another example of a 14 π electron system is II,[25] a monocyclic compound with six double bonds and one triple bond. Of course the triple bond has four π electrons, but only two of them are part of the delocalized π system; the other two form a localized π bond in the plane of the ring. Accordingly, this compound also fits Hückel's rule, and its properties indicate that it is aromatic. This is one of a number of large ring polyenes and polyenynes (i.e., compounds with both double and triple bonds) which Sondheimer and his associates have synthesized. Among others, the 18-membered ring with 9 double bonds, which Sondheimer has named [18]annulene, and the 30-membered ring [30]annulene have been made[26]; both fit Hückel's Rule. Good evidence is available that [18]annulene is aromatic, but [30]annulene is quite unstable. This high reactivity can be accounted for on a theoretical basis,[27] but it suggests that there may be a

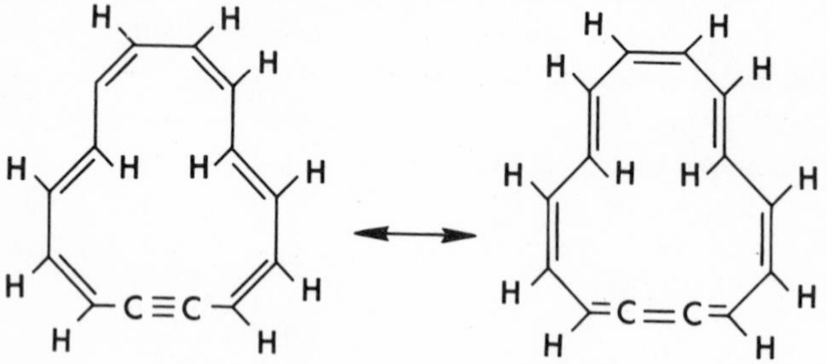

II

Dimethyl-15,16-Dihydropyrene. A Novel Aromatic System with Methyl Groups Internal to the π-Electron Cloud," *Journal of the American Chemical Society*, **85,** 1545 (1963).

25. L. Jackman, F. Sondheimer, Y. Amiel, D. Ben-Efraim, Y. Gaoni, R. Wolovsky, and A. Bothner-By, "The Nuclear Magnetic Resonance of Annulenes and Dehydro-annulenes," *Journal of the American Chemical Society*, **84,** 4307 (1962).
26. F. Sondheimer, R. Wolovsky, and Y. Amiel, "Unsaturated Macrocyclic Compounds. XXIII," *Journal of the American Chemical Society*, **84,** 274 (1962).
27. Ref. la, p. 287.

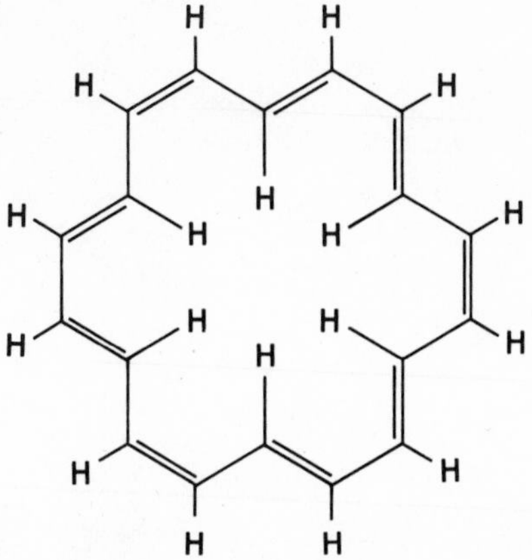

Cycloöctadecanonaene, an 18 π electron aromatic system.

practical limit to our ability to extend Hückel's rule to even larger values of *n*.

2

▪ REACTION MECHANISMS AND REACTION RATES

2–1 Classes of Organic Reactions

ORGANIC REACTIONS are conveniently classified under four headings:

1. Substitutions (sometimes called displacements)
2. Additions
3. Eliminations
4. Rearrangements

Sometimes a complex over-all reaction may fall into more than one of these categories, but the individual steps which make up the complex reaction can always be placed in one of the categories listed.

The names are self-descriptive. Thus a typical substitution reaction is the conversion of benzene to bromobenzene, substitution of a bromine for a hydrogen. Addition of bromine to a double bond, or elimination of HBr to form a double bond, exemplify the next two categories, while enolization of a ketone is a simple example of a rearrangement (rearrangements include all cases in which a change occurs only in the position of the atoms in a molecule). These classifications describe reactions, but they imply almost nothing about mechanisms.

Reactions are also classified as *heterolytic* or *homolytic* reactions. This refers to whether bonds are broken unequally, both electrons remaining with one of the atoms, or equally with one electron staying on each of the atoms.

X ⦚ :Y Heterolytic bond cleavage

X ⁝ Y Homolytic bond cleavage

Heterolytic reactions are sometimes called *ionic reactions,* while homolytic processes are involved in *free radical* reactions. Most of the organic reactions with which we will be concerned are ionic, although free radical reactions will be discussed in Chapter 7. This classification of reactions implies a bit more knowledge of the mechanisms, and there are certain cases in which it is not immediately possible to classify a reaction as homolytic or heterolytic. Thus in the pyrolysis of cyclobutane to ethylene two bonds are broken, and one could imagine either a homolytic or a heterolytic mechanism.

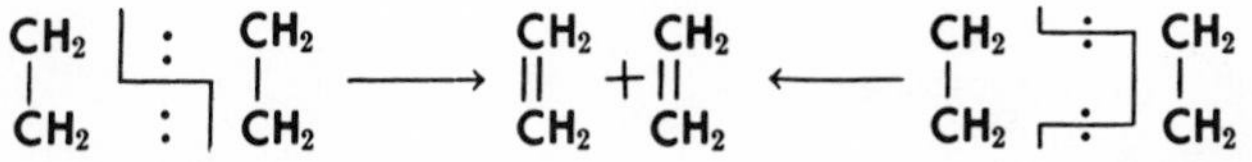

$$\begin{matrix} CH_2 & : & CH_2 \\ | & & | \\ CH_2 & : & CH_2 \end{matrix} \longrightarrow \begin{matrix} CH_2 \\ \| \\ CH_2 \end{matrix} + \begin{matrix} CH_2 \\ \| \\ CH_2 \end{matrix} \longleftarrow \begin{matrix} CH_2 & \cdot\!\cdot & CH_2 \\ | & & | \\ CH_2 & \cdot\!\cdot & CH_2 \end{matrix}$$

Heterolytic cleavage Homolytic cleavage

Because the product gives no clue to how the reaction went, and because there is no way of labeling electrons to see which one goes where, reactions of this type have sometimes been called "no-mechanism reactions." This was overly pessimistic, as we shall see in Special Topic 4.

Another classification, which is used only for the ionic reactions, depends on a particular way of describing reagents. According to the Lewis definition, a base is any species with an electron pair which may be shared, while a Lewis acid is any species which can bond to such an electron pair. Accordingly a Lewis acid is electron-seeking, while a Lewis base is nucleus-seeking. Since the terms "acid" and "base" have particular connotations, it is common to refer to reagents which attack carbon as "nucleophilic" or "electrophilic," depending on whether the reagent is an electron donor or an electron acceptor in the reaction being considered. Reactions are then classified according to the type of reagent involved. For instance, a *nucleophilic addition to a carbonyl group* occurs as one of the steps in oxime formation (see Chapter 6).

$$CH_3-\overset{\overset{\displaystyle O}{\|}}{C}-CH_3 + \ddot{N}H_2-OH \longrightarrow CH_3-\overset{O^{\ominus}}{\overset{|}{C}}-CH_3\ (\overset{\oplus}{N}H_2-OH) \left[\rightarrow \rightarrow \rightarrow CH_3-\underset{\underset{\displaystyle N\diagdown OH}{\|}}{C}-CH_3\right]$$

Here attention is called to the shifts of electron pairs by the use of curved arrows. These arrows are always used to indicate the motion of *electrons,* it being understood that the atoms will follow along. The notation is quite useful in helping to keep track of electron pairs and in calling attention to the electrophilic or nucleophilic character of any step, and we shall use it repeatedly throughout the book.

2–2 Reaction Mechanisms

For any real reaction much more detail is needed to describe the mechanism than is implied in these classifications. The mechanism is literally the detailed pathway by which the reaction occurs, and to know it would involve knowing the exact position of every atom which plays a role, in solvent molecules as well as in reacting molecules, at all times during the reaction. We would also have to know the nature of the interactions or bonds between these atoms, the energy of the system at all times, and the rate at which various changes occur during the reaction. This is more than is known for any reaction thus far, so we must select a more limited goal. For our purposes we shall consider that we have made a good start toward knowing the mechanism of a reaction if we know of all intermediate compounds that are formed, and if we can specify in general terms how each single step of the over-all reaction occurs. This general description will involve as much as we can say about which atoms are becoming attached or unattached to which others, how readily this occurs, and what sort of bonding is found as the reaction step progresses. It will be apparent throughout the book that even this limited goal is successfully achieved in very few cases. The study of reaction mechanisms is one of the most active areas of current research, and much remains to be done.

Mechanisms are generally established by excluding the reasonable alternatives and by showing that the mechanism stands up to every test which the scientist can devise. The following are important criteria in such tests.

1. The mechanism must account for the *products*. This seems trivial, but it is an important point if we include the requirement that the mechanism must account for the *stereochemistry* of the reaction and for any *isotopic labeling* results. Thus, when bromine is added to cyclohexene the product is *trans*-1,2-dibromocyclohexane, and any mechanism proposed for this reaction must account for the *trans* addition.

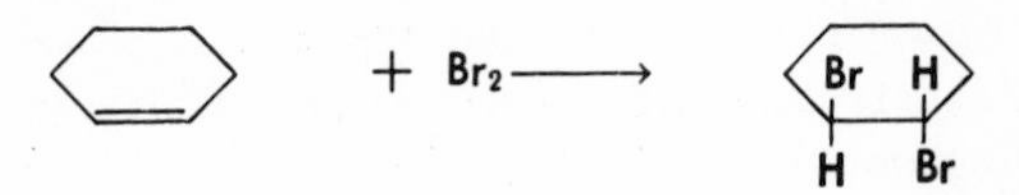

When chlorobenzene is treated with potassium amide the product is aniline, but if the starting material is labeled with radioactive carbon it has been shown that the amino group of aniline is not found only on the same carbon atom, but is on a neighboring carbon in about 50% of the aniline molecules. This result is explained by the mechanism currently in favor for this reaction (Chapter 5).

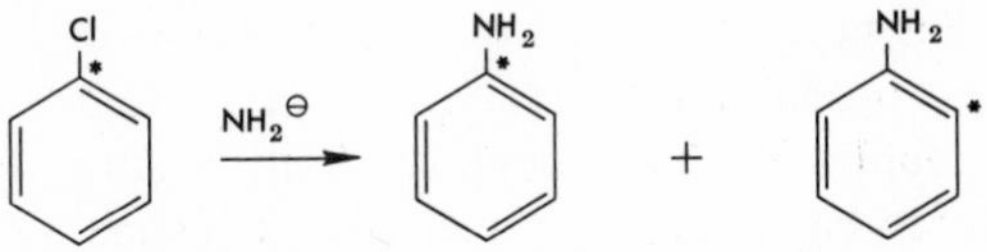

2. If *intermediates* are postulated in a mechanism it is desirable that they be detected by chemical or physical means, but in any case a real intermediate must *lead* to the correct products if it is introduced into the reaction. When this does not occur the

mechanism can be excluded. Thus when *t*-butyl bromide is allowed to stand in ethanol it produces a mixture of *t*-butyl ethyl ether and the olefin isobutylene. However, the ether is not converted to the olefin under these conditions (and vice versa); therefore the ether cannot be an intermediate in the mechanism of formation of the olefin.

$$CH_3-\underset{CH_3}{\overset{CH_3}{C}}-Br \xrightarrow{EtOH} CH_3-\overset{CH_3}{C}{=}CH_2 + CH_3-\underset{CH_3}{\overset{CH_3}{C}}-OEt + HBr$$

3. The mechanism must account for the effect of a change in reaction conditions. This includes effects on the nature of the products and on the reaction rates caused by changes in the medium or the temperature, or produced by added catalysts. It also means that the mechanism must explain the relative reactivities of related compounds. This is one of the most difficult tests to apply, since there is almost no end to the number of ways in which a reaction can be modified. This criterion will be amply illustrated throughout the book.

4. The mechanism must account for the "kinetics" of the reaction. This is really a special case of requirement 3, but it is an important one.

2–3 Kinetics

In general the rate of a reaction will depend in some way on the concentrations of reactants and catalysts. The study of this kind of dependence is part of the science of kinetics. For instance, in the reaction of hydroxide ion with methyl iodide in aqueous solution, it has been found that the rate of the reaction is proportional to the concentration of each reagent (doubling either concentration doubles the rate, doubling both quadruples the rate, etc.). This can be described by a simple rate expression:

$$OH^- + CH_3I \longrightarrow CH_3OH + I^-$$

$$\text{Rate} \equiv \frac{d[CH_3OH]}{dt} = k[OH^-][CH_3I]$$

The rate of formation of methanol, which may be expressed in differential form as shown, is proportional to the product of the concentrations of OH^- and CH_3I; the proportionality constant k

is called a *rate constant.* (It should be noted that k for this reaction is not a pure number, but that it has the units of reciprocal time and reciprocal concentration so that the units balance on both sides of the equation.)

The rate expression describes the *kinetic order* of the reaction. This reaction is of second order, since the right-hand side of the equation contains the product of two concentrations. The reaction is of first order in OH^-, since $[OH^-]$ appears to the first power.

It should be emphasized that kinetic expressions are *experimentally* derived: Within the limits of experimental error the rate of formation of methanol appears to be proportional to the OH^- and CH_3I concentrations. The kinetic order in this case is interpreted in terms of a *bimolecular mechanism.*

$$HO^{\ominus} \rightarrow CH_3 - I \longrightarrow HO - CH_3 + I^-$$

According to this mechanism OH^- simply displaces I^- from the methyl group.

The great power of kinetics is derived from an important rule: **The kinetic order of any single step of a reaction is the same as the molecularity of that step.** This is reasonable, of course, since the rate of collision between molecules A and B should depend on the concentration of each. This rule means that any suggested mechanism implies a kinetic prediction, which can be tested. Thus the bimolecular reaction of OH^- with CH_3I *must* give second-order kinetics. The observed second-order kinetics is *support* for a simple bimolecular mechanism, but other mechanisms could also show such kinetics.

An illustration of this is seen in the bromination of cyclohexene. Simple second-order kinetics is observed:

$$\text{Cyclohexene} + Br_2 \rightarrow \text{dibromocyclohexane}$$
$$\text{Rate} = k\,[\text{cyclohexene}][Br_2]$$

This would be consistent with a one-step bimolecular reaction, the direct addition of Br_2 across the double bond. However, such a reaction would lead to *cis*-dibromocyclohexane, whereas the *trans*-dibromide is actually formed. From this and other evidence (Chapter 4) it is clear that a two-step mechanism is involved:

$$\underset{\mathrm{H}\quad\quad\quad\mathrm{H}}{\diagdown\mathrm{C}{=}\mathrm{C}\diagup} + \mathrm{Br_2} \xrightarrow{\text{Slow}} \underset{\mathrm{H}\quad\quad\quad\mathrm{H}}{\diagdown\mathrm{C}\overset{\overset{\oplus}{\mathrm{Br}}}{-}\mathrm{C}\diagup}\ \ \overset{\ominus}{\mathrm{Br}} \xrightarrow{\text{Fast}} \overset{\mathrm{Br}\ \ \mathrm{H}}{\underset{\mathrm{H}\ \ \mathrm{Br}}{\diagdown\mathrm{C}{-}\mathrm{C}\diagup}}$$

The second step is fast, so the first slow step is the "bottleneck" through which the reaction must pass. This first step is bimolecular, with second-order kinetics; the kinetics of the over-all reaction is identical with that of the rate-determining step, since the rate of the reaction is the same as the rate of this slowest step.

Let us consider another reaction, the bromination of acetone in aqueous solution at constant pH (see Chapter 6). Provided there is any appreciable amount of bromine present the rate of this reaction is found to be independent of bromine concentration and first order in acetone concentration.

$$\mathrm{CH_3\text{-}CO\text{-}CH_3 + Br_2 \rightarrow CH_3\text{-}CO\text{-}CH_2Br + HBr}$$

$$\frac{d(\text{bromoacetone})}{dt} = k(\text{acetone})$$

This experimental observation is explained by the following mechanism.

$$\mathrm{CH_3-\overset{\overset{\displaystyle O}{\|}}{C}-CH_3} \xrightarrow{\quad k_1 \quad} \mathrm{CH_3-\overset{\overset{\displaystyle OH}{|}}{C}{=}CH_2}$$

$$\mathrm{CH_3-\overset{\overset{\displaystyle OH}{|}}{C}{=}CH_2 + Br_2} \xrightarrow{\ k_2\ } \mathrm{CH_3-\overset{\overset{\displaystyle O}{\|}}{C}-CH_2Br \quad + \ HBr}$$

Each simple reaction step follows the rule that the order is the same as the molecularity, so the rate of formation of acetone enol (step 1) will be first order in acetone, while the rate at which the enol is brominated will be first order in the enol as well as in bromine. The mechanism will fit the observed kinetics provided it is assumed that the second reaction is very fast compared to the first one. Then all the enol will be immediately converted to bromoacetone no matter what the bromine concentration is (unless it is vanishingly small); consequently the rate of the over-all reaction will depend only on the rate at which enol is formed. The slow step, enolization in this case, is the rate-determining step of the reaction; the subsequent faster step will not affect the kinetics of the over-all reaction since everything which gets through the bottleneck of the slow step is immediately converted to product.

At extremely low bromine concentrations the kinetics are of a different form.

$$\frac{d(\text{bromoacetone})}{dt} = k\,(\text{acetone})(Br_2) \qquad (1)$$

This is because enolization of acetone can be a reversible reaction if all the enol is not immediately trapped by bromine. Now the mechanism involves a reversible equilibrium between acetone and its enol, while bromination of the enol is rate-determining. The rate of reaction should thus be proportional to the enol and bromine concentrations.

$$\frac{d(\text{bromoacetone})}{dt} = k'(\text{enol})(Br_2)$$

However, at very low bromine concentration we are suggesting that acetone and its enol can exist in a true equilibrium, with an equilibrium constant K, so

$$K = \frac{(\text{enol})}{(\text{acetone})}; \quad (\text{enol}) = K(\text{acetone})$$

If we substitute this we get equation (1) above, provided we identify k as $k' \cdot K$. This is clearly all right, since k is just an experimentally determined proportionality constant, which may be considered to be the product of two other constants if we wish.

In addition to illustrating how the observed kinetics may change if concentrations are made drastically different, this last example points out a serious problem in interpreting reaction kinetics in terms of mechanisms. The kinetics were first order in *acetone,* but the *enol* was the species actually involved in the rate-determining step. Because the enol is in equilibrium with the acetone its concentration is proportional to that of the ketone, so the enol is said to be *kinetically equivalent* to the ketone. The observed rate law at very low bromine concentration could be interpreted as if either the acetone *or* the enol were reacting with bromine, and only other evidence shows that it is the enol which is involved. It is usually the case that more than one mechanism can be written which fits the observed kinetics of a reaction, especially when fast equilibria make several species kinetically equivalent.

2–4 Reaction Rate Theory

In our treatment so far we have written rate constants, k's, but have not discussed them further. Now we shall consider what makes one k larger than another, i.e., one reaction faster than another related one. Such information will furnish insight into the detailed mechanisms of individual steps in a reaction sequence.

Collision theory. It is possible to calculate how frequently two molecules will collide under given conditions of concentration and temperature, and it is found that in most cases only a very small fraction of such collisions results in successful reaction (often only 1 in 10^{15} collisions). To account for this it is suggested that reaction will occur only when the two particles collide with energy greater than a certain minimum amount. This minimum required energy, the *activation energy,* is needed to help break the bonds since in general the stability of the bonds being broken is lost before the new stable bonds are completely formed. The situation can be

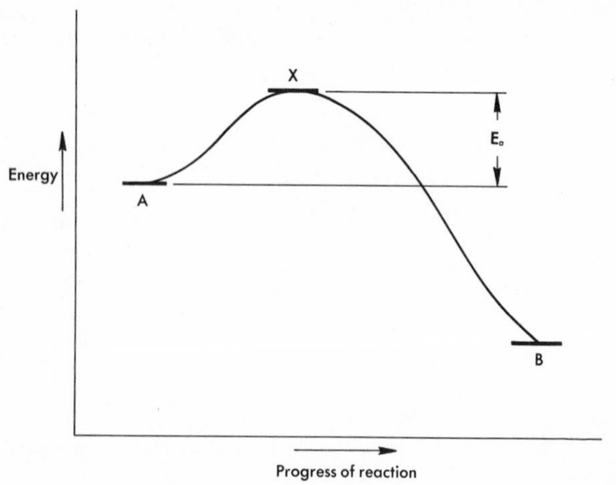

FIGURE 2–1

A collision-theory energy diagram.

pictured in terms of an energy profile for the reaction (Figure 2–1).

Thus in a simple reaction in which the starting materials (A) are less stable than the products (B) a point X will be reached during the reaction which is of higher energy than either, and which is sometimes called the "activated complex."

For instance, in the displacement reaction between OH^- and methyl iodide the activated complex occurs part-way through the reaction.

$$\underset{A}{HO^- + CH_3{-}I} \rightarrow \underset{X}{H{-}\overset{\delta-}{O}{\cdots}CH_3{\cdots}\overset{\delta-}{I}} \rightarrow \underset{B}{HO{-}CH_3 + I^-}$$

The complex, X, is not an intermediate compound but is simply an arrangement of atoms whose energy is higher than that of any other along the reaction path.

The difference between the average energy of the starting materials and the energy of this activated complex is E_a, the activation energy, and only those molecules of starting material which have this amount of energy, E_a, *more* than the average will be able to react on collision. For the reverse reaction, from B to A, the activation energy is even larger but the activated complex is the same, as it must be since there is only one "highest point" along the path which connects A and B.

It can be shown that the fraction of molecules with an energy E or greater is $e^{-E/RT}$, where e is 2.71 . . . , the base of natural logarithms. Substituting appropriate values we find that 10% of the molecules will have at least 1.4 kcal/mole more energy than average at room temperature, 1% will have at least 2.8 kcal/mole of excess energy, etc., each additional 1.4 kcal/mole causing a decrease in the fraction by a factor of 10. With this the rate of a reaction can be expressed as follows

$$\text{Rate} = Ze^{-E_a/RT}$$

where Z is the number of collisions in a unit time (for instance, per second) and the rest of the expression shows what fraction of the collisions are successful.

Some reactions obey this rate equation quite well. However, in most cases it can be shown that some collisions with the required amount of energy do not lead to reaction, because the molecules collide in the wrong way. In only a certain fraction P of the collisions will the atoms be lined up correctly for reaction even if there is sufficient energy, so this *probability factor* is usually included in the equation as well, the final form being

$$\text{Rate} = PZe^{-E_a/RT}$$

If the temperature is changed the collision frequency Z will change slightly. The major effect of an increase in temperature is to increase the fraction of molecules which have enough energy to react, so it is possible to determine the activation energy of a reaction by studying the temperature effect on the rate. The value of E_a tells us something about the strengths or energies of the various bonds in the activated complex, while the value of P tells us something about how sensitive the reaction is to the precise alignment of the reacting molecules. Thus by the application of collision theory and a study of the temperature dependence of the reaction rate it is possible to get information about the detailed mechanism of a single reaction step.

Transition state theory. There is another approach to the theory of reaction rates that is of much more use in discussing reactions in solution. Again it depends on the idea that the reacting species must surmount some energy barrier in order to react, but it discards the specific consideration of collisions, and the energy barrier is thus not formally concerned with the probability of successful collision. Instead the activation energy is considered to regulate the position of a particular equilibrium.

Consider a collision between two species which together have less than the required activation energy. By collision theory they would begin to react and form a complex which is on the way to the activated complex, but since the energy is not sufficient the reaction would not continue. Instead this complex would break up to starting materials again. If we think of this occurring repeatedly we can see that it is possible to consider an equilibrium between the starting materials and the complex; the position of such an equilibrium can be described by an equilibrium constant K.

$$A + B \overset{K}{\rightleftarrows} A \cdots B$$

Equilibrium between starting materials and a complex

All equilibrium constants can be related to the relative stabilities of starting materials and products; the relationship can be expressed in terms of the difference in their energies.

$$\Delta G° = -2.3\ RT \log K$$

Here $\Delta G°$ is the difference in *standard free energies* of products and starting materials: $\Delta G° = G°$ (products) $- G°$ (starting materials). This free energy difference is composed of two parts:

$$\Delta G° = \Delta H° - T\Delta S°$$

The *standard enthalpy difference,* $\Delta H°$, is a measure of the difference in bond energies, solvation energies, etc., between starting materials and products. The $T\Delta S°$ term multiplies the absolute temperature by a *standard entropy difference,* $\Delta S°$.

The entropy is a measure of the freedom of motion of the system, and any process which causes a decrease in freedom will cause a decrease in entropy. This factor comes into the determination of an equilibrium constant because it is relatively improbable that several freely moving particles will come together in a particular way, thus losing all their freedom to move about independently. Processes which require such loss of freedom are less favorable. All other things being equal, when two particles are bound in a complex there is an entropy decrease; therefore complexes will only be stable if the $\Delta H°$ term makes up for the $T\Delta S°$ term.

Returning to our reaction, it is postulated that the starting materials are in equilibrium with all complexes which occur before the activated complex, and *also with the activated complex itself.* Thus the concentration of the activated complex is also governed by an equilibrium constant, called $K^{\ddagger}$ to distinguish it as the constant for this particularly important equilibrium. The important postulate is made in this theory that *all activated complexes go on to products at exactly the same rate.* This can be made reasonable by the application of the theory of statistical mechanics, but it is still a remarkable postulate. If it is true, then any rate depends simply on the value of a particular equilibrium constant, $K^{\ddagger}$, for the reaction.

The universal rate constant derived from statistical mechanics is $\kappa T/h$, where κ is Boltzmann's constant, T is the absolute temperature, and h is Planck's constant; at room temperature, $\kappa T/h = 6 \times 10^{12}$ per second. All rates can be expressed in the following form:

$$\text{Rate} = \frac{\kappa T}{h}[\text{activated complex}]$$

However, the concentration of activated complex is proportional to the concentration of starting materials.

$$\frac{[\text{activated complex}]}{[\text{starting materials}]} = K^{\ddagger}$$

Accordingly, for the rate of the reaction the following expression can be written:

$$\text{Rate} = \frac{\kappa T}{h} K^{\ddagger} [\text{starting materials}]$$

This means that the rate constant is equal to $(\kappa T/h)K^{\ddagger}$. The rate constant at any particular temperature is thus proportional to the equilibrium constant $K^{\ddagger}$. The value of an equilibrium constant is determined simply by the free energy difference between starting materials and products. Thus the rate constant of a reaction depends on the standard free energy difference between starting materials and activated complex; this is written $\Delta G^{\ddagger}$, and read "free energy of activation."

$$\Delta G^{\ddagger} = -2.3\ RT \log K^{\ddagger}$$

Relative rates can be discussed in terms just like those we have already used to discuss equilibria, such as relative pK_a's of acids.

An energy diagram can be drawn for a reaction path (Figure 2–2), and in form it is almost identical to that we have already used for collision theory.

All complexes to the left of the activated complex are considered to be in equilibrium with the starting materials, while all complexes

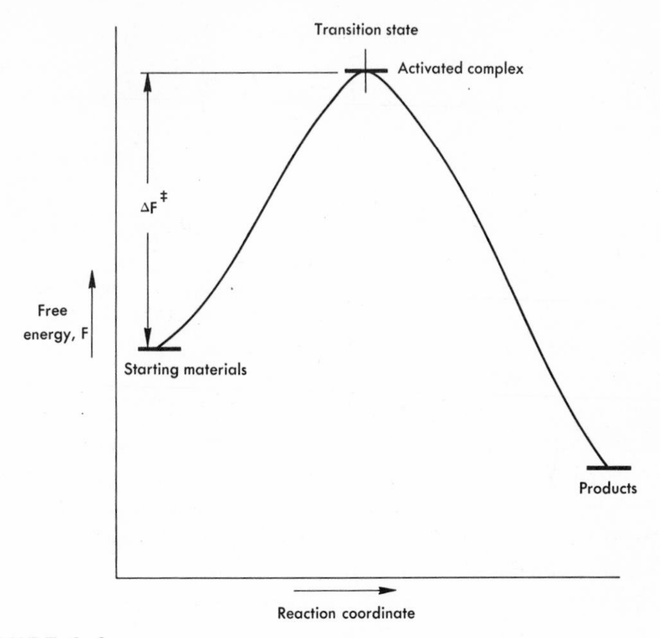

FIGURE 2–2

A transition-state theory energy diagram.

to the right are considered to be in equilibrium with the products. Thus the activated complex occurs at what is called the *transition state* (the place where a transition occurs from the starting-material side to the product side); quite commonly the activated complex is actually called the transition state.

It is a simple matter to represent a complex series of reactions on a single energy diagram. For instance, when benzaldehyde is treated with potassium cyanide in ethanol, benzoin is formed by the following mechanism (cf. Chapter 6):

$$C_6H_5\overset{\displaystyle H}{\overset{|}{C}}{=}O + {}^{-}C{\equiv}N \underset{k_{-1}}{\overset{k_1}{\rightleftarrows}} \underset{\displaystyle \mathbf{I}}{C_6H_5\overset{\displaystyle H}{\overset{|}{\underset{\underset{\displaystyle N}{\overset{|||}{\displaystyle C}}}{\underset{|}{C}}}}{-}O^{-}} \underset{k_{-2}}{\overset{k_2}{\rightleftarrows}} \underset{\displaystyle \mathbf{II}}{C_6H_5{-}\overset{-}{\underset{\underset{\displaystyle N}{\overset{|||}{\displaystyle C}}}{\underset{|}{C}}}{-}OH}$$

$$\text{II} + C_6H_5\overset{H}{\overset{|}{C}}{=}O \xrightarrow{k_3} C_6H_5-\underset{\underset{C\equiv N}{|}}{\overset{OH}{\overset{|}{C}}}-\underset{\underset{H}{|}}{\overset{{}^{-}O}{\overset{|}{C}}}-C_6H_5 \xrightarrow{k_4}$$

$$\text{III}$$

$$C_6H_5-\underset{\underset{C\equiv N}{|}}{\overset{O^{\ominus}}{\overset{|}{C}}}-\underset{\underset{H}{|}}{\overset{HO}{\overset{|}{C}}}-C_6H_5 \xrightarrow{k_5} C_6H_5\overset{O}{\overset{\|}{C}}-\underset{\underset{H}{|}}{\overset{OH}{\overset{|}{C}}}-C_6H_5 + {}^{-}C\equiv N$$

IV Benzoin

Here five reactions have been written, two equilibria and three irreversible reactions (although, of course, no reaction is truly irreversible, for all practical purposes the reversibility of all steps after the slowest one can usually be ignored). Each of these individual reactions will have its own transition state. The highest point of energy in the entire sequence will occur at the transition state for the rate-determining step, which is in this case step 3, with rate constant k_3. Thus the rate of formation of benzoin will depend *only* on the difference in free energy between the starting materials, benzaldehyde and cyanide, and the activated complex at the transition state of step 3. This is shown in Figure 2–3, in which the $\Delta G^{\ddagger}$ for the overall sequence is shown rather than that for any given step.

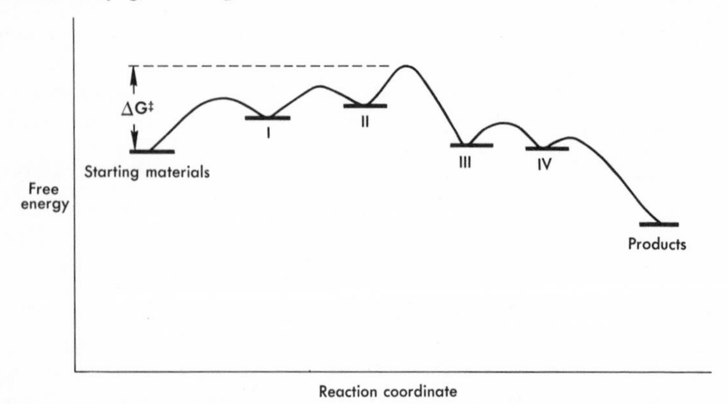

FIGURE 2–3

An energy diagram for the benzoin condensation.

This means that the rate of the reaction can simply be described in terms of an equilibrium between starting materials and activated complex, all other equilibria or reaction steps being ignored.

$$2\ C_6H_5CHO + CN^{\ominus} \underset{}{\overset{K^{\ddagger}}{\rightleftharpoons}} \left[C_6H_5-\overset{\overset{OH}{|}}{\underset{\underset{C\equiv N}{|}}{C}}\text{-----}\overset{\overset{O}{\vdots|}}{\underset{\underset{H}{|}}{C}}-C_6H_5 \right]^{\ominus}$$

Starting materials **Activated complex**

Since the rate will depend only on the concentration of activated complex this equilibrium tells us immediately that the kinetics will be of the following form (as is found experimentally).

$$\text{Rate} = k\left[C_6H_5\overset{\overset{OH}{|}}{\underset{\underset{CN}{|}}{C}}\ \ominus\right]\left[C_6H_5CHO\right] = k'\left[C_6H_5CHO\right]^2\left[CN^-\right]$$

Note that the first expression is kinetically equivalent to the second. Furthermore, in considering what effect a change of temperature, pH, solvent, etc., has on the reaction rate we must consider its effect only on this equilibrium, and ignore the effect on individual steps. This makes the interpretation simpler, but it points out once again that kinetics will not give us information about fast equilibria which occur before the rate-determining step.

2–5 Catalysis

A catalyst is a substance which accelerates a reaction without being itself changed in the over-all process. It must do this, of course, either by lowering the transition state energy of the uncatalyzed process in some way or by making a new path possible which has a low activation energy. Many catalysts function in ways which are still not fully understood; in particular, heterogeneous catalysts (those not in solution) such as the platinum catalyst in hydrogenations are in this category. On the other hand, definite mechanisms can be suggested for the action of most catalysts in solution; the operation of cyanide ion in the benzoin mechanism above is a good example of this. Since by far the most important cases of homogeneous catalysis in organic chemistry involve the action of acids and bases, they will be discussed in a little detail.

General acid and general base catalysis. A solution of acetic acid in water contains two acidic species, H_3O^+ and HOAc (ignoring H_2O, a very weak acid). The concentration of each of these can be measured independently, since the pH is a measure only of the H_3O^+, whereas a titration would indicate the total acid present. Some reaction rates depend only on the pH; such a reaction whose rate is increased at a lower pH is said to be *specific acid catalyzed,* since only the solvent-related acid H_3O^+ plays a catalytic role. By contrast, some reactions may show *general acid catalysis;* the rate will increase if the HOAc concentration is increased even if the pH is held constant.

When the bromination of acetone is conducted in water solution of varying pH's and acetic acid concentrations, it is found that the reaction is catalyzed by both acidic species, and a simple rate law is observed.

$$\text{Rate} = k_1\,[\text{acetone}][H_3O^+] + k_2\,[\text{acetone}][\text{HOAc}]$$

The reaction is studied with enough bromine present to guarantee that enolization is rate-determining; therefore bromine does not appear in the rate expression.

If we consider what is involved in the enolization of acetone it is easy to see why an acid catalyst would help. A proton must be removed from carbon and another proton added to oxygen; if only water is present the solvent must perform these functions.

$$\underset{\substack{|\\ H\\ H_2O:}}{CH_3-\overset{\substack{H-OH\\ O\\ \|}}{C}-CH_2} \longrightarrow CH_3-\overset{\substack{H\\ /\\ O\\ |}}{C}=CH_2 \qquad {}^{\ominus}OH \qquad \overset{\oplus}{H-OH_2}$$

However, it would clearly be better if the proton could be donated by a stronger acid than water, and removed by a stronger base. Thus the function of an acid in catalyzing this enolization can be written very simply.

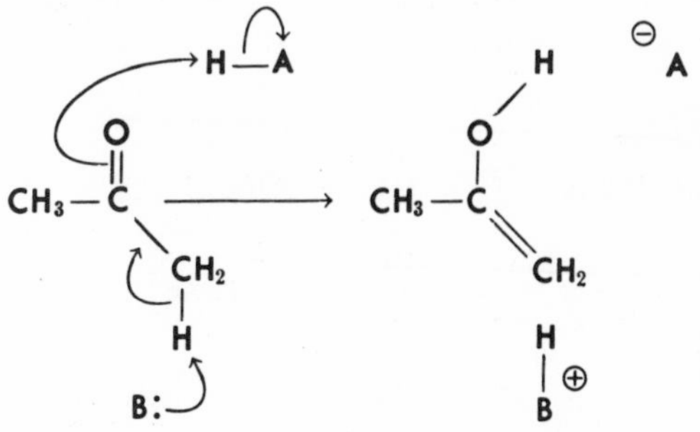

$$CH_3-\overset{\displaystyle O}{\overset{\|}{C}}-CH_2-H + H-A + B: \longrightarrow CH_3-\overset{\displaystyle OH}{\overset{|}{C}}=CH_2 + A^{\ominus} + H-B^{\oplus}$$

Here we have suggested not only that the proton donor may be any acidic species, but also that the proton acceptor may be any base. This is also found to be the case. For instance, in a solution of sodium acetate both OH^- and OAc^- catalyze the bromination reaction, and their contribution to the rate can also be described with a simple kinetic expression.

$$\text{Rate} = k_3\,[\text{acetone}][OH^-] + k_4\,[\text{acetone}][OAc^-]$$

If both acids and bases are present, as in a buffer solution of acetic acid and sodium acetate in water, the complete kinetic expression is the following.

$$\begin{aligned}\text{Rate} = {} & k_0\,[\text{acetone}] + k_1\,[\text{acetone}][H_3O^+] + \\ & k_2\,[\text{acetone}][HOAc] + k_3\,[\text{acetone}][OH^-] + \\ & k_4\,[\text{acetone}][OAc^-] + \quad k_5\,[\text{acetone}][HOAc][OAc^-]\end{aligned}$$

The first term represents enolization accomplished simply by water. (Water does not appear in the kinetic expression since it is the solvent whose concentration is kept constant; only concentrations which can be experimentally varied are shown, since the kinetics are based on the *observed* variation of rate with concentration.) The second term represents enolization in which H_3O^+ is the proton donor and H_2O the base, etc. The sixth term indicates that acetate ion can remove the proton from carbon while acetic acid is putting it on the oxygen, and this raises an interesting point. Why is there no term for a reaction in which OAc^- and H_3O^+ cooperate, or HOAc and OH^-? The answer is that these terms are already hidden in the expression we have written. In water, acetic acid is in equilibrium with H_3O^+ and OAc^-.

$$HOAc \underset{}{\overset{K}{\rightleftarrows}} H_3O^{\oplus} + OAc^{\ominus}$$

$$[H_3O^{\oplus}][OAc^{\ominus}] = K\,[HOAc]$$

$$\therefore\; k[H_3O^{\oplus}][OAc^{\ominus}] = k'\,[HOAc]$$

This equilibrium shows that the third term in the kinetic expression does not necessarily represent catalysis by acetic acid, but it may actually indicate catalysis by H_3O^+ and OAc^-, since the equilibrium makes them kinetically equivalent. Similarly, the fifth term, which indicates catalysis by OAc^- and water, may really indicate catalysis by HOAc and OH^-, since this combination is again kinetically equivalent. Other evidence suggests that both types of catalysis contribute to the terms in the observed kinetic expression.

This example indicates again that it is usually easier to measure kinetics than to interpret them unambiguously. Even so, it seems clear that acid or base catalysis must be accounted for in terms of definite molecular mechanisms, and not in terms of mysterious catalytic "forces." In later chapters, we shall encounter other examples of acid and base catalysis.

The Brønsted catalysis law. In the previous example we have seen that a proton can be removed from acetone by OH^-, OAc^-, or H_2O. Of course, one might expect that reactions involving strongly basic OH^- should be faster than those with the less basic OAc^- or H_2O, but this expectation is only intuitive. "Basicity" is an equilibrium concept, involving starting materials and products, while rates of reactions involve transition state energies.

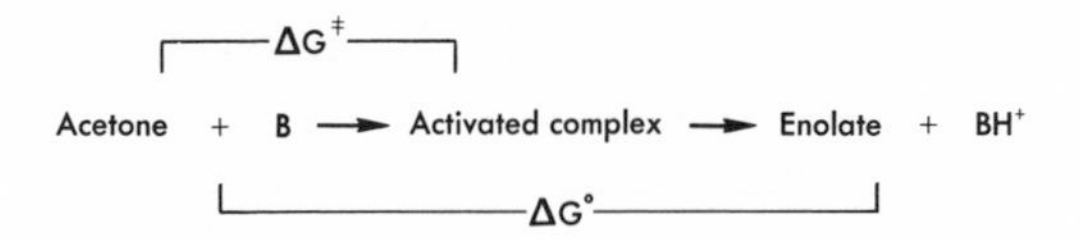

With a stronger base ΔG° will become more negative, since the conversion from B to BH^+ becomes more favorable, but there is no obvious relationship between ΔG° and $\Delta G^\ddagger$. Interestingly, it is frequently found experimentally that with bases of similar structure (e.g., carboxylate anions) but differing basicities the changes in ΔG° and $\Delta G^\ddagger$ are related:

> The change in $\Delta G^\ddagger$ from one base to another is proportional to the change in ΔG°, or $\Delta(\Delta G^\ddagger) = k\Delta(\Delta G^\circ)$.

This is reasonable, since in the activated complex a proton has been partially transferred to the base, and we might expect to see some fraction of the effect of basicity which shows up fully when the transfer is complete.

The foregoing *linear free energy relationship* can be expressed in more standard form by the Brønsted relation:

$$k_B = G_B K_B^{\beta}$$

This expression indicates that the rate constant k_B for a base catalyzed reaction is proportional to the dissociation constant K_B for the base, raised to a power β. The exponent β is called the Brønsted coefficient; it varies in magnitude between 0 and 1, and its value reflects the extent to which proton transfer has occurred in the transition state and thus the extent to which the strength of the base can affect the activation energy. G_B is a simple proportionality constant.

In the case of acetone enolization, a series of carboxylate anions has been examined as catalysts and the Brønsted law is followed, with $\beta = 0.88$. This value so close to 1 indicates that almost the full effect of the basicity is already felt at the transition state; an increase of 10 in the strength of the base leads to an increase of 7.5 (or $10^{0.88}$) in the rate. In cases where the transition state occurs earlier β is smaller in value. There is also a Brønsted relationship for acid-catalyzed reactions:

$$K_A = G_A K_A^{\alpha}$$

and it also is generally found to hold reasonably well among a series of acids with similar steric requirements but differing acid strengths.

General References

A. Frost and R. Pearson, *Kinetics and Mechanism* (2nd ed., John Wiley & Sons, New York, 1961). The best general reference for this chapter. Covers kinetics, rate theory, and catalysis; in Chapter 11, several reaction mechanisms are discussed in detail.

R. P. Bell, *The Proton in Chemistry* (Cornell University Press, Ithaca, New York, 1959). Particularly useful for material on acid and base catalysis.

S. L. Friess, E. S. Lewis, and A. Weissberger, eds., *Investigation of Rates and Mechanisms of Reactions* (2nd ed., in *Technique of Organic Chemistry,* Vol. VIII, Interscience Publishers, New York, 1961). An exhaustive treatment of the methods used to examine reaction mechanisms. Discusses both the experimental details and the mechanistic implications of such techniques as isotopic labeling, kinetics, detection of intermediates, etc.

2

Special Topic

▪ CATALYSIS BY ENZYMES

ESSENTIALLY ALL BIOLOGICAL reactions are catalyzed by special substances called enzymes.[1] These catalysts are proteins, which sometimes are associated with inorganic ions or small organic molecules; enzymes are so large that it has been possible to determine the three-dimensional structure of only a few of them, although the study of protein structures is now a very active field of research.[2] In general, each reaction requires a different kind of enzyme, so that there are almost as many types of catalyst molecules as there are biological reactions. Thus there are groups of enzymes which catalyze various oxidation reactions, other groups which catalyze hydrolyses, others which catalyze carbon-carbon bond formation, etc. This means that in detailed mechanism each enzyme-catalyzed reaction is a special case, and a difficult one

1. For more extensive treatments of this topic see (a) S. Bernhard, *The Structure and Function of Enzymes* (W. A. Benjamin, Inc., New York, 1968); (b) T. Bruice and S. Benkovic, *Bioorganic Mechanisms* (W. A. Benjamin, Inc., New York, 1966).
2. The current state of this field can be followed in *Annual Review of Biochemistry, Advances in Enzymology,* or *Advances in Protein Chemistry.*

considering the enormous size and complexity of the catalyst. Even so, the general principles by which enzymes work can be simply understood in terms of reaction-rate theory.

There are several respects in which catalysis by these substances differs from that due to simple catalytic species such as H_3O^+, CN^-, etc. One of the most striking is that enzymes are generally *much better catalysts* than are simple molecules. For instance, at 25°C the hydrolytic enzyme α-chymotrypsin[3] (molecular weight ~ 25,000) will catalyze the hydrolysis of an amide to a carboxylic acid and an amine; as a catalyst it is about one million times more effective[4] than either H_3O^+ or OH^-, although the enzyme operates in neutral solution. The second important feature is that enzymes are *very selective* in their action. Chymotrypsin, for instance, will not hydrolyze amides in general, but will operate only on amides of certain amino acids. The selectivity of enzymes is so great that most will work with only one enantiomer of a DL pair in an optically active substrate. This specificity is related to the requirement that the substrate have just the right shape to fit into a particular place in the protein, for the third special feature of enzyme-catalyzed reactions is that the enzyme binds the substrate in a *complex* before it catalyzes any reaction.

Thus enzyme-catalyzed reactions can be written in a very general way,[5] in which the enzyme is symbolized by E, the substrate by S, and the enzyme-substrate complex by ES.

$$E + S \overset{K}{\rightleftarrows} ES \overset{k}{\rightarrow} E + \text{products}$$

General scheme for enzyme-catalyzed reactions

This complexing can be detected by spectroscopic means, but it was first deduced from a study of the kinetics of enzyme-catalyzed

3. P. Desnuelle, "Chymotrypsin," in P. Boyer, H. Lardy, and K. Myrbäck, eds., *The Enzymes*, Vol. 4 (Academic Press, New York, 1960). Chapter 5.
4. T. Bruice, "Intramolecular Catalysis and the Mechanism of Chymotrypsin Action," *Brookhaven Symposia in Biology*, **15**, 52 (1962).
5. Cf. H. Mahler and E. Cordes, *Basic Biological Chemistry* (Harper & Row, New York, 1968).

reactions.[6] A typical graph of rate versus substrate concentration is shown in Figure 2–4. At low concentrations of substrate the rate follows a normal law,

$$\text{Rate} = k\ (\text{enzyme})\ (\text{substrate})$$

but at high substrate concentration the rate depends only on the total enzyme concentration

$$\text{Rate} = k\ (\text{total enzyme})$$

and is independent of substrate concentration.

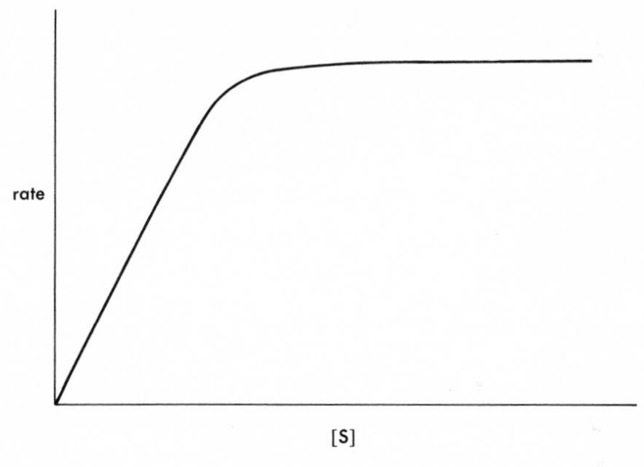

FIGURE 2–4

Typical plot of rate versus substrate concentration for an enzyme-catalyzed reaction.

This result comes about because the rate-determining step is the further reaction of the enzyme-substrate complex, and the over-all rate of the reaction will thus depend simply on the concentration of ES and on the rate constant for its further reaction.

$$\text{Rate} = k\ (\text{ES})$$

Now enzyme-substrate complexes are generally quite stable; with a reasonably high substrate concentration it is possible to push the

6. L. Michaelis and M. Menten, "Die Kinetic der Invertinwirkung," *Biochemische Zeitschrift*, **49**, 333 (1913).

equilibrium to the right so that essentially all the enzyme is tied up in the complex. This is the maximum concentration the complex can have, so it is not surprising that higher concentrations of substrate do not increase the rate further. A mathematical treatment[5] of this general mechanism, in which the rate reaches a maximum value as all the enzyme becomes complexed, predicts the exact shape of the plot of rate versus concentration, and the fit with experimental results is very good.

A number of interactions are responsible for the formation of such complexes.[7] These include (1) attractions between charged groups on the protein and the substrate, (2) hydrogen bonds between suitable groups, and (3) "hydrophobic interaction," the tendency of nonpolar groups such as alkyl chains to come together rather than be surrounded by water molecules (in a sense the hydrocarbon sections of a protein "extract" nonpolar substrates out of the water). The total of such binding forces can be considerable. Thus in the complexing between chymotrypsin and benzoyltyrosylamide, a typical substrate,

$$C_6H_5C(=O)-NH-CH(CH_2C_6H_4OH)-C(=O)-NH_2$$

Benzoyltyrosylamide, a substrate for chymotrypsin

the equilibrium constant[8] indicates a $\Delta G°$ of -2 kcal/mole on going to the complex; such an energy decrease indicates that complexing is a favorable process. However, from the temperature effect on the equilibrium it can be calculated that the $\Delta H°$ for complex formation is -11 kcal/mole; the binding forces cause

7. (a) R. Lumry, "Some Aspects of the Thermodynamics and Mechanism of Enzymic Catalysis," and (b) K. Linderstrøm-Lang and J. Schellman, "Protein Structure and Enzyme Activity," in P. Boyer, H. Lardy, and K. Myrbäck, eds., *The Enzymes,* Vol. 1 (Academic Press, New York, 1959). Chapters 4 and 10.
8. Ref. 7a, p. 178.

this large decrease in energy. The $-T\Delta S°$ term is $+9$ kcal/mole, showing that complex formation is very unfavorable in entropy terms. This is reasonable, since there is considerable loss of freedom when the two independent particles, substrate and protein, are tied into one. Thus the binding forces make it possible for a stable complex to form in spite of the unfavorable entropy change.

Such enzyme-substrate complexes are formed in all cases, although of course the exact energetics of the process depends on the particular system being considered. Once in the complex, the substrate is acted on by catalytic groups within the protein, usually side-chain groups of amino acids which are part of the protein. Again the detailed mechanism depends on the particular reaction being considered. In the case of chymotrypsin, the following mechanism seems likely.[9] A general base, the imidazole ring of a histidine, removes the proton from a serine hydroxyl group as the latter attacks the amide carbonyl. The departing amine may be simultaneously protonated by a general acid group, although direct evidence on this point is not yet available. In a second step water attacks the new ester carbonyl group, perhaps assisted by a general base, while the departing serine oxygen is simultaneously protonated by imidazolium ion (Figure 2–5).

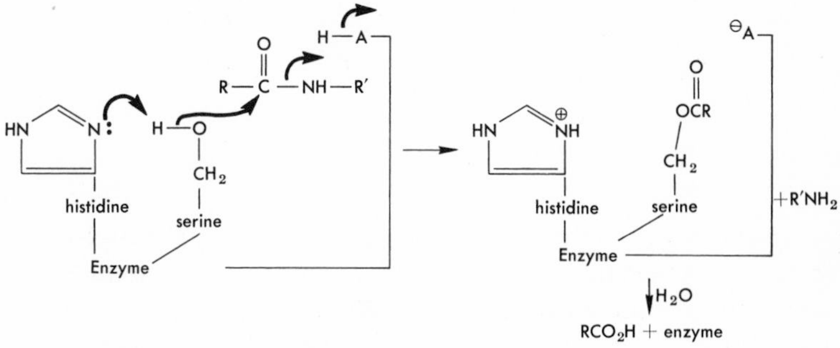

FIGURE 2–5

A general scheme for amide hydrolysis by α-chymotrypsin.

9. M. Bender *et al.*, series of papers on the mechanism of action of α-chymotrypsin, *Journal of the American Chemical Society*, **86**, 3704, 5330 (1964) and references therein. H. Neurath, "Protein Digesting Enzymes," *Scientific American*, December 1964, p. 68.

The enzyme ribonuclease operates in quite a similar way.[1] This enzyme is a small protein, made up of 124 amino acids, which catalyzes the hydrolysis of ribonucleic acids. As with chymotrypsin the hydrolysis is a two-step process: The leaving group is first displaced by an alcoholic hydroxyl group, and water comes into the second step in the hydrolysis of the intermediate. In the case of ribonuclease, however, the nucleophilic hydroxyl is on the substrate, ribonucleic acid, rather than on the enzyme as in the chymotrypsin reaction.

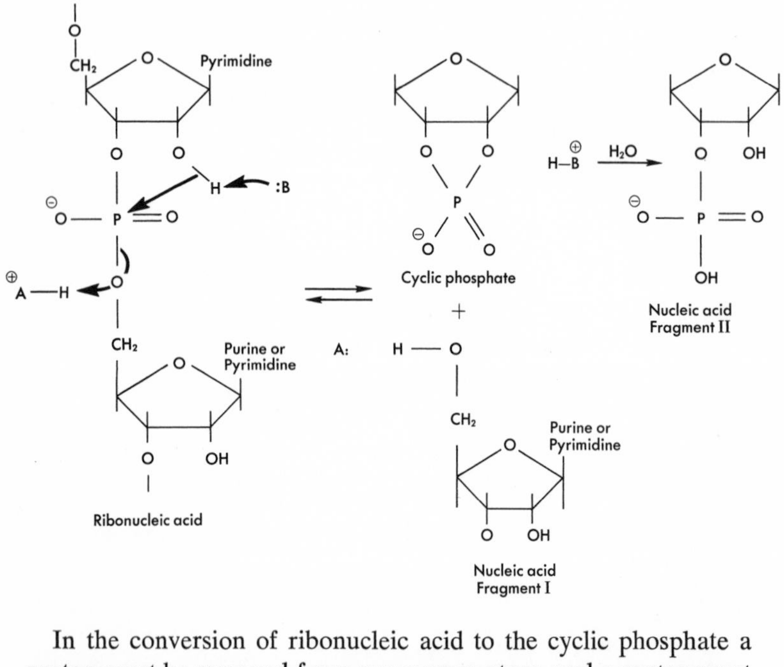

Ribonucleic acid

Cyclic phosphate

Nucleic acid Fragment I

Nucleic acid Fragment II

In the conversion of ribonucleic acid to the cyclic phosphate a proton must be removed from one oxygen atom, and a proton must be added to another oxygen atom. We have shown a base B: removing the first proton and an acid AH^+ adding the other; in the reverse reaction, in which Fragment I adds to the cyclic phosphate to produce the nucleic acid, the law of microscopic reversibility dictates that we must run back over the same path, so the new base A: will remove the proton from Fragment I while the new acid BH^+ will act as proton donor. Although the detailed mechanisms of enzyme-catalyzed reactions have not been completely elucidated in any case, strong evidence indicates that B: is the imidazole of histidine-119 (the polypeptide chain is numbered from the end with a free α-amino group), and AH^+ is a protonated imidazole of histidine-12. Although one might at first expect the two histidines to have the same pK_a, the difference in energy between B: and BH^+ is affected by the environment (neighboring charges, for instance) so His-12 and His-119 may exist as shown, with 12 protonated and 119 as the free base. In the reverse reaction A: is thus again an imidazole ring, while BH^+, the proton donor for the reverse reaction, is again an imidazolium ion. In step 2, the hydrolysis of the cyclic phosphate, this same mechanism is involved, but water replaces Fragment I, so A: is removing the proton from H_2O in the transition state.

Lysine (1) --------- Histidine (12) -------- Histidine (119) ------- Valine (124)

H_2N — CH — CO ---- NH — CH — CO ---- NH — CH — CO ---- NH — CH — CO_2H

R (Lys-1); CH_2–imidazolium (NH, NH ⊕) (His-12); CH_2–imidazole (NH, N) (His-119); R (Val-124)

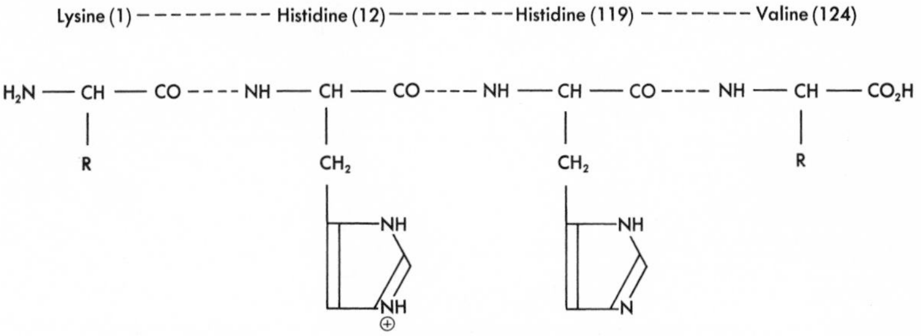

Ribonuclease

Both chymotrypsin and ribonuclease accomplish hydrolysis by a two-step sequence, in which a leaving group is set free before a water molecule is tied down. It is easy to see that this is catalytically useful when we realize that the free energy of activation $\Delta G^‡$ involves both an enthalpy and an entropy term; the latter, $\Delta S^‡$, measures the loss of degrees of freedom in the transition state compared with starting materials—the more freedom lost the larger $\Delta G^‡$, and the slower the reaction. With a two-step hydrolysis mechanism either step could be rate determining. If the first one is, then H_2O is not yet tied down in the transition state for over-all reaction, so its loss of freedom does not contribute to $\Delta S^‡$; if the second step is rate-determining, then the H_2O is tied down, but the leaving group is already fully free, so its gain in degrees of freedom (compared with starting materials) balances the ΔS from the water. Either situation is better than a one-step process in which water is tied down before the leaving group is fully released.

The other point of interest is that imidazole was used as the base by both enzymes, and imidazolium ion as an acid by ribonuclease. This feature is also catalytically useful. Imidazole, with a pK_a near 7, is the strongest base which can exist free at neutrality; if it were more basic, it would be already protonated by the medium at pH 7. Similarly, imidazolium ion is the strongest acid available at neutrality, since an acid with a lower pK_a would be dissociated at pH 7. By the Brønsted relationships the rate of an acid- or base-catalyzed reaction depends on the acid or base strength of the catalyst, so imidazole is optimal for these functions.

Clearly an important part of the catalytic efficiency of any enzyme must be due to ideal rigid placement of such groups so that they can perform their functions without any further restriction of freedom (loss of entropy). We shall not consider this point further, but shall examine only one additional question: How does enzyme-substrate complex formation contribute to the catalytic efficiency of enzymes?[10]

In transition-state theory the rate of a reaction depends on the change of free energy in going from starting materials to activated complex. We have emphasized that the free energy change depends on two factors: $\Delta H^\ddagger$, which reflects changes in potential energy, and $\Delta S^\ddagger$, which reflects changes in freedom. For a catalyst these two will ordinarily oppose each other. Thus if we wish to stabilize a developing negative charge in the transition state for a reaction, we may do this with a general acid which donates a proton as the charge develops. This would lower $\Delta H^\ddagger$, and thus help the reaction; but it would also decrease $\Delta S^\ddagger$, since the general acid would be tied down next to the substrate at the transition state and freedom would be lost. Since $\Delta G^\ddagger = \Delta H^\ddagger - T\Delta S^\ddagger$, the entropy loss works against the enthalpy decrease. If we require several independent catalyst molecules, e.g., a general base, a general acid, a nucleophile, etc., the entropy loss becomes very large; for some possible catalysts, this entropy loss may be more than the possible enthalpy gain, so catalysis (rate increase) does not occur.

The entropy problem can be partly solved if all the catalytic groups are placed in the same molecule to start with. This is the case for an enzyme; now the translational freedom of only one molecule is lost, as contrasted with the situation when the catalytic

10. Cf. F. Westheimer, "Mechanisms Related to Enzyme Catalysis," *Advances in Enzymology,* **24,** 455 (1962).

groups are in independent molecules. However, there is still an important entropy decrease when an enzyme and its substrate go to the activated complex. The two must be tied together, with a loss of translational freedom, and they must be fixed in precisely the right orientation, so that some freedom of rotation within the molecules is lost as well. The enzyme balances this entropy loss with an *extra* enthalpy decrease, the enthalpy decrease involved in *binding* to form the complex.

The situation is pictured in Figure 2–6. With ordinary cata-

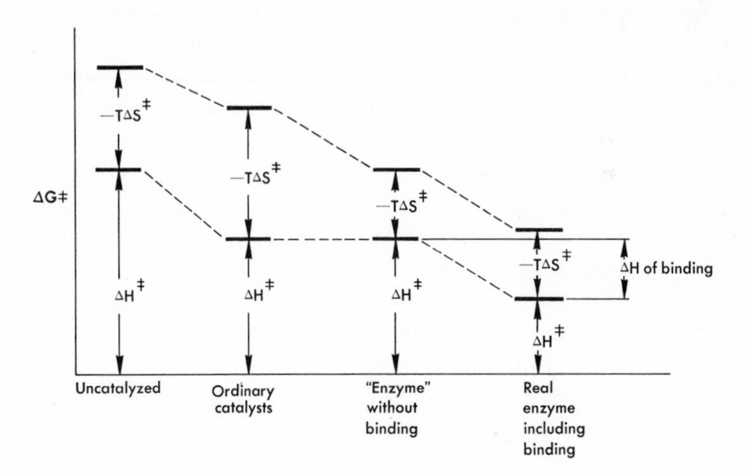

FIGURE 2–6

The effect of various types of catalysts on $\Delta H^{\ddagger}$ *and* $T\Delta S^{\ddagger}$.

lysts the rate can be increased because of an enthalpy advantage, which is, however, opposed by the entropy disadvantage discussed above. If one imagined an "enzyme" which could not bind substrate there would still be a big entropy advantage in combining all the catalytic groups in one molecule (for simplicity we have assumed that $\Delta H^{\ddagger}$ does not change since the same catalytic groups have been used). Finally the real enzyme has the extra feature that the activated complex is bound to the enzyme, so the enthalpy of binding is subtracted as well. As we mentioned earlier, for chymotrypsin with a particular substrate this enthalpy of binding was −11 kcal/mole. Since each 1.4 kcal/mole corresponds to a factor of 10 in the rate, the binding should result in a rate increase of 10^8, or 100,000,000.

Several questions may be raised about the preceding analysis. First of all, the measured enthalpy of binding is that for the substrate, while we are really interested in the enthalpy of binding of the activated complex. This may not be the same, and there are even suggestions that it may be greater,[11] but the magnitude is probably comparable to that measured for the substrate. Second, it might be objected that both the substrate and the activated complex are bound, so the energy *difference* between them is unaffected by binding. However, the enzyme is ordinarily present in only catalytic amounts; therefore the true starting materials for any measured reaction (when concentrations are such that the enzyme is not largely complexed by either starting materials or products) are the free substrate and enzyme, not the ES complex. The great virtue of transition-state theory is that it focuses attention on the difference in energy between starting materials and activated complex, and makes it clear that various intermediates between the two are not relevant. In this sense the fact that an enzyme-substrate complex forms is not strictly relevant, since this is an intermediate. Only the argument above that binding of the activated complex is probably similar makes enzyme-substrate binding forces of interest.

11. Ref. 7a, p. 222.

3

NUCLEOPHILIC ALIPHATIC SUBSTITUTION

OF ALL CLASSES OF REACTIONS, substitution at saturated carbon atoms by nucleophilic reagents has been studied most thoroughly. In this chapter we shall first consider the evidence for the currently accepted reaction mechanisms; the last section of the chapter is concerned with the relationship between structure and reactivity.

3–1 Reaction Mechanisms

S_N2 reactions. The reaction of methyl bromide with sodium hydroxide solution, to afford methanol, is first order in methyl bromide and first order in hydroxide ion. Accordingly, both species are present in the transition state of the reaction, which has a simple direct-displacement mechanism. This mechanism is described by Ingold by the symbol S_N2, meaning *substitution nucleophilic bimolecular*. It should be noted that the 2 stands for bimolecular, not for second order; the order of a reaction is a real *experimental* quantity, while molecularity refers to the number of species involved (undergoing a change in covalency) in the rate-determining step of a particular *mechanism* being considered.

$$HO^{\ominus} \; CH_3—Br \longrightarrow CH_3—OH + Br^{\ominus}$$

$$\text{Rate} = k\,[CH_3Br][OH^-]$$

It is found experimentally that second-order kinetics, implying an S_N2 mechanism, are found for a variety of displacement reactions.

When the carbon atom on which displacement occurs is asymmetrically substituted it is possible to investigate the *stereochemistry* of the reaction, and in all cases examined it is found that displacement occurs with inversion, so the nucleophile begins bonding to the back of the carbon atom while the leaving group is departing from the front. Accordingly the process can be visualized as is shown below.

$$Y^{\ominus} \;\; R_1R_2R_3C—X \longrightarrow Y\text{----}C(R_1R_2R_3)\text{----}X \longrightarrow Y—CR_1R_2R_3 + X^{\ominus}$$

Starting materials **Transition state** **Products**

An S_N2 displacement

This inversion of configuration during S_N2 reactions is not due simply to electrostatic repulsion between an attacking hydroxide ion and a leaving halide ion, for instance; even displacements in which there should be electrostatic *attraction* between the nucleophile and the leaving group proceed with inversion. A good example is found in the reaction of D-α-phenethyltrimethylammonium ion with acetate ion by an S_N2 mechanism. In the transition state there should be attraction between the attacking acetate group, which still has some negative charge, and the departing trimethylamine, which still has some positive charge. Even so, the reaction proceeds with complete inversion.

$$CH_3-C\begin{matrix}\diagup\!\!\diagup O \\ \diagdown O^-\end{matrix} \;+\; H\cdots C(C_6H_5)(CH_3)-\overset{\oplus}{N}(CH_3)_3 \longrightarrow$$

$$CH_3-C\begin{matrix}\diagup\!\!\diagup O \\ \diagdown \overset{\ominus}{O}\end{matrix}\cdots\cdots C(C_6H_5)(H)(CH_3)\cdots\cdots\overset{\oplus}{N}(CH_3)_3 \longrightarrow$$

$$CH_3-C\begin{matrix}\diagup\!\!\diagup O \\ \diagdown O\end{matrix}-C(C_6H_5)(CH_3)\cdots H \;+\; :N(CH_3)_3$$

In molecular orbital terms the S_N2 displacement can be described as follows. The nucleophile approaches the sp^3 orbital which is being used by carbon to bond the displaceable group, and the nucleophile begins to overlap with the smaller lobe of this orbital. As bonding proceeds the carbon rehybridizes so that eventually both the nucleophile and the leaving group are bonded to a carbon p orbital, one to each lobe. The process continues as carbon goes from this sp^2 hybrid state on to the product, in which it is again sp^3 hybridized. The fact that inversion is always observed simply reflects the better bonding in this type of transition state. The other possibility is not as good, overlap of both nucleophile and electrophile with the same lobe of an sp^3 orbital in a front-side displacement.

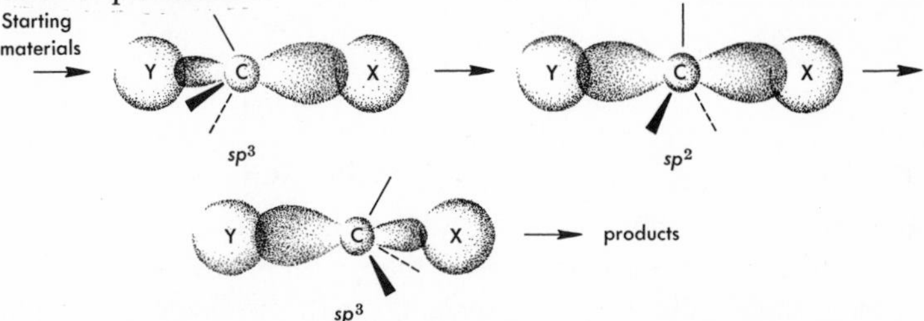

S_N1 reactions. In the S_N1 mechanism (substitution nucleophilic unimolecular) substitution occurs in two steps. First the leaving group ionizes, leaving a carbonium ion, and then the carbonium ion reacts with the nuceophile. Carbonium ions are so unstable that they will react very rapidly with any nucleophile (except for a few exceptional carbonium ions, such as tropylium ion); thus in general the first step, ionization, will be rate-determining. Accordingly the kinetics should be first-order, the rate depending on the concentration of substrate but not on that of the nucleophile.

$$R - X \xrightarrow{\text{slow}} R^+ + X^-$$

$$R^+ + Y^- \xrightarrow{\text{fast}} R - Y$$

$$\text{Rate} = k\,[RX]$$

Such kinetic behavior is often observed for nucleophilic substitutions when an intermediate carbonium ion can be stabilized by resonance or inductive effects. For instance, benzhydryl chloride reacts with fluoride ion by an S_N1 mechanism in liquid SO_2 solution, for the rate of the reaction does not depend on the concentration of fluoride ion (in the concentration range studied).

$$C_6H_5CH(Cl)-C_6H_5 + F^- \xrightarrow[\text{solution}]{SO_2} C_6H_5-CH(F)-C_6H_5 + Cl^-$$

$$\text{Rate} = k\,[\text{benzhydryl chloride}]$$

$$C_6H_5-CH(Cl)-C_6H_5 + C_5H_5N \xrightarrow{SO_2} C_6H_5-CH(N^{\oplus}C_5H_5)-C_6H_5 \quad Cl^{\ominus}$$

$$C_6H_5-CH(Cl)-C_6H_5 + Et_3N \xrightarrow{SO_2} C_6H_5-CH(^{\oplus}NEt_3)-C_6H_5 \quad Cl^{\ominus}$$

The reaction rate does not depend on the concentration of the nucleophile, nor does it depend on the *nature* of the nucleophile. If benzhydryl chloride is reacted with pyridine or with triethylamine instead of with fluoride ion the rate of reaction is the *same,* although the products are of course different; again the kinetics is found to be first-order, independent of nucleophile concentration.

The kinetic picture is not quite so simple as we have indicated for these substitutions. As the reaction proceeds it is found that the chloride ion which is produced can also act as nucleophile, so it begins to compete for the carbonium ion. Accordingly, the rate at which substitution occurs can be decreased by the addition of chloride ion, which traps some of the reactive intermediate and returns it to starting material. This is an example of the *common ion effect,* and it is characteristic of an S_N1 reaction; we can take it into account by indicating a reversible ionization in our mechanism.

$$C_6H_5-\underset{}{\overset{Cl}{\overset{|}{C}H}}-C_6H_5 \underset{k_{-1}}{\overset{k_1}{\rightleftarrows}} C_6H_5-\underset{\oplus}{\overset{H}{\overset{|}{C}}}-C_6H_5 + Cl^{\ominus} \xrightarrow[\text{Nucleophile}]{k_2} \text{Product}$$

The fraction of intermediate ion which goes on to products is simply the ratio of two rates, involving rate constants and concentrations.

$$\text{Fraction of carbonium ion converted to products} = \frac{k_2(\text{nucleophile})}{k_{-1}(Cl^-) + k_2(\text{nucleophile})}$$

Thus the rate of substitution, taking the common ion effect into consideration, is really not independent of the concentration of nucleophile

$$\text{Rate} = k_1(\text{R-Cl})\ \frac{k_2(\text{nucleophile})}{k_{-1}(Cl^-) + k_2(\text{nucleophile})}$$

Experimentally, first-order kinetic behavior can still be found by examining the early part of the reaction before much common ion has been formed, but the common ion effect can also be of use in detecting an S_N1 mechanism. One example of this is discussed below.

Solvolysis. A simple kinetic test to decide between the S_N1 and the S_N2 mechanisms fails if the nucleophile is a solvent molecule. The determination experimentally of whether or not the nucleophile is involved in the rate-determining step requires that we be able to vary the concentration of the nucleophile, which is not possible for the solvent. For instance, the hydrolysis of *t*-butyl bromide in water is independent of the concentration of hydroxide ion.

$$\text{t} - \text{Bu} - \text{Br} + \text{OH}^{\ominus} \xrightarrow{H_2O} \text{tBu} - \text{OH} + \text{Br}^{\ominus}$$

$$\text{Rate} = k\,[\text{t} - \text{Bu} - \text{Br}]$$

This shows that the mechanism does not involve an S_N2 displacement by hydroxide ion, but it does not rule out S_N2 displacement by water.

$$\text{t} - \text{BuBr} + H_2O \xrightarrow{S_N2?} \text{t} - \text{Bu}\overset{+}{\text{O}}H_2 + \text{Br}^{\ominus} \underset{\text{fast}}{\xrightarrow{OH^-}} \text{t} - \text{BuOH} + H_2O$$

It might be thought that this problem could be solved by going to mixed solvents, so the concentration of the water could indeed be varied. When aqueous acetone is the solvent it is found that the rate of hydrolysis of *t*-butyl bromide increases if the solvent is changed from 10 to 30% water, but the increase is 40-fold for this 3-fold change in concentration. This is mostly because the reaction is faster in the more polar solvent, but it is not possible to decide whether the full 40-fold increase is due to the increased polarity of the medium. This polarity change might contribute a factor of 13 to the rate, and the remaining factor of 3 could be present because water appears in the kinetic expression as a nucleophile. Accordingly, it is not possible to use kinetics in this way to determine whether the solvent is acting as a nucleophile in an S_N2 process or whether instead an S_N1 mechanism is involved.

In some cases the common ion effect can be used to solve this problem. Thus when benzhydryl chloride is hydrolyzed in aqueous acetone the rate is unaffected by base, but this does not remove the ambiguity we have just discussed, the choice between an S_N1 mechanism and an S_N2 displacement by water. However, it is found that while the addition of many salts increases the rate, since they make the medium more polar, the addition of lithium *chloride* markedly slows the hydrolysis. The common ion effect is unambiguous evidence that this hydrolysis has an S_N1

mechanism. However, the common ion effect will only be seen if the carbonium ion intermediate is stable enough that it can survive in a sea of solvent molecules until it can be trapped by the common ion. Unfortunately, the *t*-butyl cation is relatively reactive; there is a small decrease in the rate of hydrolysis of *t*-butyl bromide caused by added bromide ion, but the effect is so slight that it really is not good evidence for an S_N1 mechanism.

Perhaps the most convincing evidence for an S_N1 mechanism in a solvolysis reaction is the *stereochemistry* of the reaction. As we have seen, S_N2 reactions proceed with inversion of configuration at the carbon being substituted. A carbonium ion is flat, the carbon being sp^2 hybridized and using the three hybrid orbitals for single bonding; the remaining *p* orbital is empty. Accordingly, once a carbonium ion is formed it is equally likely to be attacked on either side (i.e., on either lobe of the *p* orbital); therefore, an S_N1 mechanism should lead to *racemization* rather than preferential inversion or retention of configuration.

$$R'R''RC{-}X \longrightarrow R'R''RC^{\oplus} \xrightarrow{Y^-} R'R''RC{-}Y + Y{-}CRR'R''$$

This test has been applied to solvolysis. When optically active α-phenylethylchloride is submitted to S_N2 conditions, i.e., treatment with the very reactive sodium methoxide or sodium ethoxide, one does observe second-order kinetics and clean inversion, as expected.

$$C_6H_5{-}\underset{}{\overset{Cl}{\overset{|}{C}H}}{-}CH_3 + NaOR \xrightarrow{ROH} C_6H_5{-}\overset{OR}{\overset{|}{C}H}{-}CH_3 \text{ with inversion}$$

$$\text{Rate} = k[R{-}Cl][OR^-]$$

S_N2 reaction with inversion

Under solvolysis conditions, e.g., in aqueous acetone, the solvolysis indeed has an S_N1 mechanism.

$$C_6H_5{-}\overset{Cl}{\overset{|}{C}H}{-}CH_3 + H_2O \xrightarrow[\text{Acetone}]{\text{80\% Aqueous}} C_6H_5{-}\overset{OH}{\overset{|}{C}H}{-}CH_3$$

98% Racemized

S_N1 reaction with racemization

However, the "real life" result is almost always more complicated than simple theories predict, and it is important to note that racemization is not complete but is accompanied by 2% of net inversion. Such partial inversion can be more extensive in other systems. Thus solvolysis of α-phenylethyl chloride in acetic acid yields the acetate which is only 85% racemized; the remaining 15% has an inverted configuration, showing that the acetic acid has added from the side opposite to the leaving group.

$$C_6H_5-\underset{\displaystyle}{\overset{\displaystyle Cl}{\overset{|}{C}H}}-CH_3 + CH_3\overset{\displaystyle O}{\overset{\|}{C}}OH \longrightarrow C_6H_5-\overset{\displaystyle O\overset{\displaystyle O}{\overset{\|}{C}}CH_3}{\overset{|}{C}H}-CH_3 + HCl$$

85% Racemized

15% Inverted

Ion pairs in $\boldsymbol{S_N1}$ **reactions.** Such observations, racemization accompanied by some inversion, are very common. They might be explained by the suggestion that there are two simultaneous independent mechanisms, S_N1 reaction and some S_N2 displacement by solvent at the same time, but several facts make this explanation unattractive. For instance, the same racemization with partial inversion is observed in the solvolysis of some tertiary alkyl halides or esters, although other evidence (vide infra) shows that such compounds are very unreactive in ordinary S_N2 displacements because of steric hindrance at the carbon atom.

$$CH_3-\overset{\displaystyle CH_3}{\overset{|}{C}H}-CH_2-\underset{\displaystyle \underset{\displaystyle \underset{\displaystyle C_6H_4CO_2H}{|}}{\overset{|}{C}=O}}{\underset{|}{\overset{\displaystyle CH_3}{\overset{|}{C}}}}\! \underset{\displaystyle O}{} -CH_2-CH_3 \xrightarrow{CH_3OH} CH_3-\overset{\displaystyle CH_3}{\overset{|}{C}H}-CH_2-\underset{\displaystyle OCH_3}{\underset{|}{\overset{\displaystyle CH_3}{\overset{|}{C}}}}-CH_2-CH_3 + o\text{-}C_6H_4(CO_2H)(CO_2H)$$

40% Racemized

60% Inverted

Instead of invoking S_N2 displacement, inversion is explained by considering that the leaving group is still near the carbonium ion when reaction occurs. Thus it is suggested that the ionization of an alkyl halide leads first to an *ion pair,* in which the halide ion helps solvate the carbonium ion on one side. Solvation on the other side by an ordinary solvent molecule is also expected. If this solvated species collapses to product, there will be inversion, while racemization can occur if the halide ion is first replaced by a normal solvent species, leading to a symmetrically solvated ion.

$$RR'R''C\text{—}X \xrightarrow{H_2O} H_2O\text{:}\cdots \overset{+}{C}RR'R''\cdots \bar{X} \longrightarrow HO\text{—}CRR'R'' \quad \text{(Inversion)}$$

$$H_2O\text{:}\cdots \overset{+}{C}RR'R''\cdots \bar{X} \xrightarrow{H_2O\text{:}} H_2O\text{:}\cdots \overset{+}{C}RR'R''\cdots \text{:}OH_2 \longrightarrow HO\text{—}CRR'R'' + RR'R''C\text{—}OH \quad \text{(Racemization)}$$

The solvated ions can also be represented in molecular orbital terms if "solvation" is considered to involve weak overlap of orbitals.

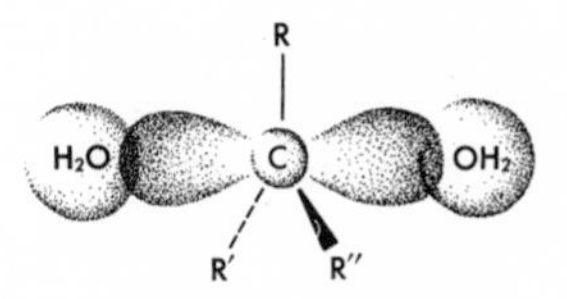

The reader will have noticed that when a single solvent molecule is specifically written in coordination with one lobe of the *p* orbital, the picture is almost identical with that previously drawn for the S_N2 transition state. The difference is that in the S_N2 case the entering and leaving groups are so strongly coordinated with the central carbon atom that there is no chance for the leaving group to be replaced by another nucleophile before collapse to products; thus a symmetrical species cannot be formed, and inversion occurs. This strong coordination also means that the central carbon atom has no appreciable positive charge. In the solvated carbonium ion, the intermediate has enough lifetime that the leaving group can be exchanged for another solvent molecule, and the coordination is sufficiently weak that in many respects the carbon can be considered to have a full positive charge. However, when the solvent molecules are specifically considered in this picture of the S_N1 mechanism, it is apparent that the S_N1 and S_N2 mechanisms are very closely related. The mechanisms of real reactions may well be intermediate between them, in some cases.

Reactions of allylic halides. Allylic halides react readily by an S_N1 mechanism, since the intermediate carbonium ion is resonance stabilized.

$$R{-}CH{=}CH{-}CH_2{-}Cl \longrightarrow RCH{=}CH{-}\overset{\oplus}{C}H_2 \longleftrightarrow R\overset{\oplus}{C}H{-}CH{=}CH_2$$

$$\xrightarrow{H_2O} R{-}CH{=}CH{-}CH_2OH + R{-}\underset{}{\overset{OH}{\overset{|}{C}H}}{-}CH{=}CH_2$$

As would be expected from this mechanism, the intermediate carbonium ion can react at either positive center, and a mixture of allylic isomers is generally obtained. Of course the same carbonium ion could be formed by starting with the isomeric chloride, so it is possible to test this mechanism by examining the composition of the product mixture obtained from each starting material. When this is done the usual observation is made—real life results are more complicated than simple theories predict. Thus, in comparing the reactions of crotyl chloride, $CH_3CH{=}CH{-}CH_2{-}Cl$, with those of its allylic isomer α-methylallyl chloride, $CH_3{-}CHCl{-}CH{=}CH_2$, one finds that on solvolysis in ethanol each leads to a mixture of products, but the composition of the mixture is not quite the same for each.

	$CH_3CH{=}CH{-}CH_2OEt$	+ $CH_3\overset{OEt}{\overset{|}{C}H}{-}CH{=}CH_2$
$CH_3{-}CH{=}CH{-}CH_2{-}Cl \xrightarrow[78°C]{EtOH}$	92%	8%
$CH_3{-}\overset{Cl}{\overset{|}{C}H}{-}CH{=}CH_2 \xrightarrow[78°C]{EtOH}$	82%	18%

This result can be explained by proposing that simultaneous S_N1 and S_N2 reactions with solvent are taking place. Alternatively, as the chloride ionizes, solvation may take place first on the carbon atom from which the halogen is leaving; this solvated cation could either collapse to a product of the same allylic structure, or go on to a more symmetrically solvated allylic cation which can lead to both isomers.

A very important observation on the solvolysis of allylic halides adds strong support to the idea that a tight ion pair of carbonium ion and leaving group is the first intermediate. Thus when α,α-dimethylallyl chloride (I) is solvolyzed in acetic acid it undergoes rearrangement to the allylic isomer (II) simultaneous with acetolysis. This rearrangement is a first-order process, independent of added chloride ion, so it does not involve some sort of allylic displacement reaction. More important, when radioactive chloride ion is added to the solution the rearranged product has only a fraction of the expected radioactivity. This shows that most of the chlorine atoms in the product are directly derived from the chlorine of the starting material, and not from free chloride ions. The only attractive explanation of these facts is that the starting material ionizes and that the tight ion pair formed rearranges and collapses to the isomeric chloride before it can react with external chloride ions.

$$\underset{\text{I}}{Cl-C(CH_3)_2-CH=CH_2} \xrightarrow{HOAc} \underset{\text{Solvent}}{\overset{\bar{Cl}\cdots HOAc}{(CH_3)_2\overset{+}{C}-CH=CH_2}} \rightarrow$$

$$\underset{\text{Solvent}}{\overset{AcOH\cdots\cdots \bar{Cl}}{(CH_3)_2C=CH-\overset{+}{C}H_2}} \longrightarrow \underset{\text{II}}{(CH_3)_2C=CH-CH_2Cl}$$

This phenomenon, collapse of the initially formed ion pair which is detected because the structure has rearranged, has been observed in the solvolysis of many allylic compounds; it is usually called "internal return." In all these cases it can be shown that the particular chloride ion (or other leaving group ion) formed by dissociation is preferentially recaptured in the rearranged product, supporting the idea that this particular ion is still closely associated with the carbonium ion.

Although these allylic rearrangements were first-order processes which did not involve any external nucleophile kinetically, it might be wondered whether an S_N2-like reaction *could* occur with rearrangement in allylic compounds. Such a process, called the S_N2' reaction, has indeed been observed in some cases, although it is not common. The first example was found in the reaction of α-ethylallyl chloride with the sodium enolate of malonic ester. The kinetics are strictly second order, but 23% of the product was of rearranged structure, so an S_N2' reaction accompanies the normal S_N2 process.

$$CH_3-CH_2-CH(Cl)-CH=CH_2 + {}^{\ominus}CH(COEt{=}O)_2 \quad Na^{\oplus} \longrightarrow$$

$$\underset{77\%}{CH_3-CH_2-CH(CH(CO_2Et)_2)-CH=CH_2} + \underset{23\%}{CH_3CH_2-CH=CH-CH_2-CH(CO_2Et)_2}$$

$$\text{Rate} = k[\text{R—Cl}][\text{enolate ion}]$$

A study of the stereochemistry of the S_N2' reaction was performed by treating the substituted cyclohexenyl dichlorobenzoate (III) with piperidine. The kinetics were cleanly second-order, and the product was entirely the rearranged isomer (IV) resulting from S_N2' rather than S_N2 attack. Interestingly, it was found that the entering group comes in entirely from the side of the ring that holds the leaving group.

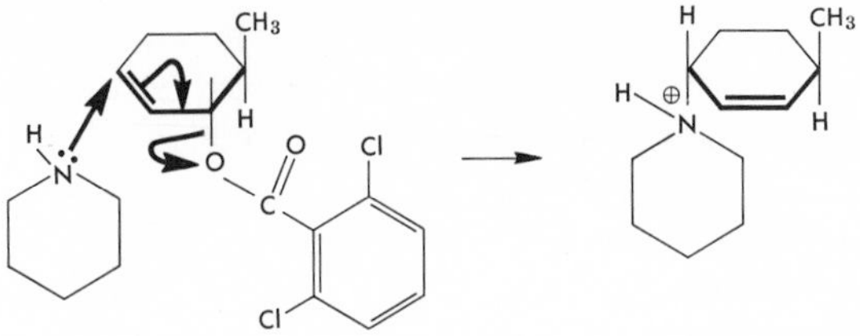

III $\longrightarrow$ IV

The chemistry of allylic compounds shows that the nucleophilic substitution reactions of an alkyl halide are strongly affected by the presence of an adjacent double bond. Special effects are found for many other neighboring groups as well, and these will be considered in the next section.

Neighboring group participation. If optically active 2-bromopropionic acid is treated with dilute methanolic sodium methoxide, substitution of bromine by a methoxyl group takes place in a reaction with first-order kinetics (i.e., with no dependence on the concentration of methoxide ion). It is found that this substitution has occurred with complete *retention* of configuration at carbon. Thus one has an apparent S_N1 reaction in which there is not the usual racemization with partial inversion, but instead retention of stereochemistry, a result different from that expected for either the S_N1 or the S_N2 mechanisms. Such results are best explained by invoking a double displacement mechanism. First there is attack on carbon by the neighboring carboxylate ion to form an α-lactone, and then displacement occurs by the methoxide ion to form the final products. If both reactions occur with inversion the result will be over-all retention of configuration, as observed.

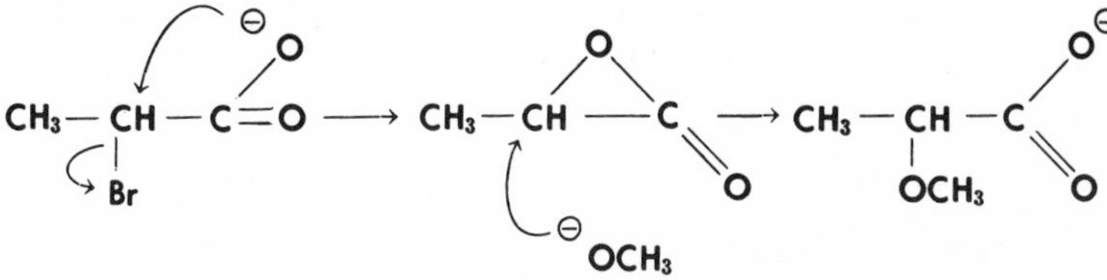

$$CH_3-CH(Br)-C(=O)O^{\ominus} \longrightarrow CH_3-CH\overset{O}{\frown}C{=}O \xrightarrow{{}^{\ominus}OCH_3} CH_3-CH(OCH_3)-C(=O)O^{\ominus}$$

The kinetics show that the first step is rate-determining, methoxide ion playing a role only after the transition state and thus not appearing in the rate expression. The α-lactone is highly strained, and is more reactive than an ordinary ester would be.

Groups other than carboxylate ion can participate in neighboring group reactions. A well-known example is the reaction of ethylene chlorohydrin with sodium ethoxide, to form ethylene oxide.

$$HO-CH_2-CH_2Cl + NaOEt \longrightarrow {}^{\ominus}O-CH_2-CH_2-Cl \longrightarrow CH_2(O)CH_2 + Cl^{\ominus}$$

In this case internal displacement leads to a stable product (although with more vigorous treatment the oxide may be opened by a second displacement). Not only does the reaction proceed only with internal displacement, rather than with S_N2 substitution by ethoxide, but the reaction of ethylene chlorohydrin with sodium ethoxide is 5100 times faster than a comparable S_N2 reaction, displacement on ethyl chloride by sodium ethoxide. This large preference for intramolecular reaction, even at the expense of forming a strained ring, is due to probability factors. In collision-theory terms, the nucleophile is permanently held next to the carbon it must attack, so reaction can occur whenever the species picks up enough energy. In transition-state terms, an ordinary S_N2 reaction involves a loss of entropy when the nucleophile and substrate are tied down in the transition state. In an internal reaction there is no need to tie down a second molecule, so the entropy of activation is much more favorable. This can even make up for the unfavorable enthalpy associated with making a strained ring.

An illustration of such factors is found in the internal displacement reactions of aminoalkyl halides.

$$Br-CH_2-(CH_2)_n-NH_2 \longrightarrow \overline{CH_2-(CH_2)_n-\overset{+}{N}H_2} \quad Br^-$$

As Table 3–1 shows, formation of the strainless five-membered ring is fastest. The also strainless six-membered ring is formed more slowly, since on the average the two reacting groups are

TABLE 3–1 ▪

Rates of Cyclization of Aminoalkyl Bromides in H_2O, 25° C

rate = k (substrate)

Substrate	k (per second)
Br—CH_2—CH_2—NH_2	6×10^{-4}
Br—CH_2—CH_2—CH_2—NH_2	0.08×10^{-4}
Br—CH_2—CH_2—CH_2—CH_2—NH_2	5000×10^{-4}
Br—CH_2—CH_2—CH_2—CH_2—CH_2—NH_2	80×10^{-4}
Br—CH_2—CH_2—CH_2—CH_2—CH_2—CH_2—NH_2	0.1×10^{-4}

further apart and have less probability of reacting. Thus as the ring to be formed is made larger the advantage of internal reaction becomes less, since it becomes less probable that the two ends of the chain will collide. Both the three-membered and four-membered rings are strained and formed more slowly, but probability factors make the three-membered ring formation faster than the four even though the smaller ring is more strained.

Mustard gas, Cl—CH_2—CH_2—S—CH_2—CH_2—Cl, is a very reactive alkylating agent toward all nucleophiles, and this property is responsible for its vesicant action on the skin, since it alkylates proteins. The rate of hydrolysis is independent of hydroxide ion, supporting a two-step process, and the hydrolysis is slowed by chloride ion. These data indicate participation of the neighboring sulfur.

$$ClCH_2CH_2—\ddot{S}—CH_2CH_2—Cl \rightleftharpoons Cl—CH_2CH_2—\overset{\oplus}{S}\langle(CH_2—CH_2)\ \ Cl^{\ominus}$$

$$\xrightarrow{H_2O} Cl—CH_2—CH_2SCH_2CH_2OH$$

Further evidence for the formation of a cyclic sulfonium ion in a related case is the observation that both V and VI are transformed to the same chloride VII on treatment with HCl. This is expected if the isomeric starting alcohols are transformed to a common intermediate, the cyclic sulfonium ion.

$CH_3CH_2SCH(CH_3)—CH_2OH$ (V) $\xrightarrow{HCl}$ $CH_3CH_2—\overset{\oplus}{S}\langle$ $CH(CH_3)$, CH_2 (three-membered ring) $+ Cl^{\ominus}$ $\rightarrow$ $CH_3CH_2—S—CH_2CH(Cl)—CH_3$ (VII)

$CH_3CH_2SCH_2CH(CH_3)—OH$ (VI) $\xrightarrow{HCl}$ $CH_3CH_2—\overset{\oplus}{S}\langle$ $CH(CH_3)$, CH_2 (three-membered ring)

V

VI

VII

Evidence of the types cited is available now for neighboring group participation by alkoxy groups, ester groups, halogen atoms, phenyl groups, and a variety of others. Some of this evidence will be further discussed in Special Topic 3.

3–2 Reactivity in Nucleophilic Substitution

S_N2 reactions

The Structure of the Alkyl Group. The relative reactivity of methyl chloride versus ethyl chloride in an S_N2 displacement depends on the nature of the nucleophile, solvent, and reaction conditions. However, Streitwieser has analyzed the mass of data available in the literature and has come up with a set of "average" relative reactivities for various alkyl derivatives in S_N2 reactions. These are listed in Table 3–2.

TABLE 3–2 ▪

Average Relative Rates of Alkyl Systems in S_N2 Reactions

Alkyl Group	Relative Rate
Methyl	30
Ethyl	1
n-Propyl	0.4
n-Butyl	0.4
Isopropyl	0.025
Isobutyl	0.03
t-Butyl	Nil
Neopentyl	0.00001
Allyl	40
Benzyl	120

Steric hindrance clearly plays a major role, so methyl compounds are the most reactive of the simple alkyl derivatives. The largest steric effect comes when alkyl groups are substituted directly at the carbon to be attacked. Thus there is a large effect on going from methyl to ethyl to isopropyl to *t*-butyl, but only a minor effect on going from ethyl to *n*-propyl. One exception to this generalization is found when the carbon *next* to the reaction site is highly branched, for then it can severely hinder approach to its neighbor; this effect is seen in the very low reactivity of neopentyl compounds.

CH_3
C — CH_2
CH_3 CH_3 X

Steric shielding in a neopentyl derivative.

Electronic effects also play a role, as the high reactivity of allyl and benzyl derivatives shows. At the transition state for the S_N2 reaction the carbon being substituted is sp^2 hybridized, and uses its other *p* orbital to bond the entering and leaving groups. In the allyl and benzyl compounds this *p* orbital can also be conjugated with the rest of the π electron system, explaining the greater stability of these transition states and thus the higher reaction rates.

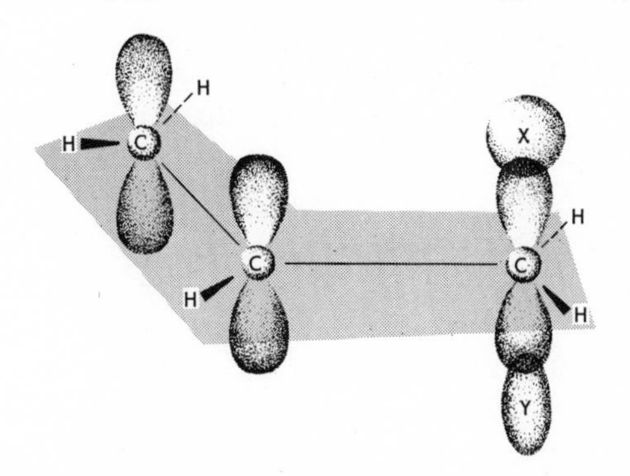

Transition state for an S_N2 substitution in an allyl compound.

High reactivity due to electronic effects is also found in α-haloketones; for instance, chloroacetone is 33,000 times as reactive as is *n*-propyl chloride towards potassium iodide (in acetone at 50°).

$I^{\ominus}$ + $CH_3-\underset{\underset{O}{\|}}{C}-CH_2-Cl$ $\qquad$ $I^{\ominus}$ + $CH_3-CH_2-CH_2-Cl$

$k_{relative}$: 33,000 $\qquad$ 1.0

In this case the energy of the transition state is lowered by some bonding between the nucleophile and the carbonyl group.

Very interesting steric effects are found in some cyclic compounds. In Table 3–3 are listed the rate constants for S_N2 reaction of iodide ion with some cycloalkyl bromides.

TABLE 3–3 ■

S_N2 Reactivities of Some Cycloalkyl Bromides toward I^- in Acetone at 70°C

Alkyl Group	k (liters/mole second) × 10^7
Cyclopropyl	< 0.01
Cyclobutyl	0.98
Cyclopentyl	208

It is apparent that the three- and four-membered ring compounds are much less reactive than the cyclopentyl. Since the transition state for an S_N2 reaction has sp^2 hybridized carbon while the starting material has sp^3 hybridized carbon, the bond angles at carbon should go from 109°28′ in the starting material to 120° in the transition state. In cyclopropyl bromide the bond angle is only 60°. This is 49°28′ smaller than sp^3 hybridization requires, so the molecule has considerable strain (called I-strain, or "internal" strain); in the transition state the strain would be even worse, the angle being 60° less than sp^2 hybridization requires. Accordingly I-strain raises the energy of the transition state more than that of the ground state. The activation energy thus becomes larger, and therefore the rate is slower. The same effect, to a lesser degree, is found for cyclobutyl bromide. This type of strain effect is only important for the smaller rings; in five, six, and larger sized rings more subtle effects due to steric hindrance by neighboring hydrogens can be detected.

The Nature of the Nucleophile. The relative reactivities of different nucleophiles will depend on substrate, conditions, etc. Again a table of "average" nucleophilicities has been assembled by Streitwieser, and selected values are shown in Table 3–4.

At first one might have expected that this order would reflect the basicity of the nucleophiles. In the transition state the nucleophile is beginning to form a bond to carbon; the stronger that bond is the lower the energy of the transition state should be, and thus the faster the reaction (although basicity measures the tendency to bond to hydrogen, and carbon "basicity" need not parallel hydrogen basicity). However, it is clear from a glance at the table that such basic species as phenoxide or acetate ions are less nucleophilic than a nonbasic ion such as I^-. Several factors are involved in determining nucleophilicity. First of all, when the attacking atom is the same, e.g., oxygen, basicity does play a role,

TABLE 3–4 ■

Relative Nucleophilicities

Nucleophile	Relative Rate
$C_6H_5S^-$	470,000
I^-	3,700
EtO^-	1,000
Br^-	500
$C_6H_5O^-$	400
Cl^-	80
CH_3COO^-	20
Pyridine	20
NO_3^-	1

so one finds the order: ethoxide > phenoxide > acetate > nitrate. Secondly, larger atoms are more nucleophilic than smaller ones. This shows up in the enormous reactivity of thiophenoxide ion compared to phenoxide, even though the latter is more basic. The relative reactivities of the halide ions also reflect this factor. The size effect has been partly attributed to *polarizability:* the electrons of a large atom are less firmly held by the nucleus, so they may more easily move in response to some demand. In the transition state for S_N2 reaction the nucleophile is still at a large distance from carbon; the transition state will be stabilized if the electrons of a polarizable atom can shift toward the carbon so as to allow effective bonding even at this great distance.

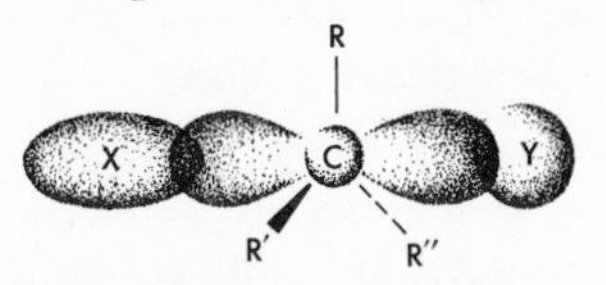

Attack by a polarizable nucleophile, X.

Larger atoms also tend to hydrogen bond poorly with solvent hydroxyls, so fewer bonds to the solvent need be broken during the substitution reaction. This factor means that relative nucleophilicities may be very different in nonhydroxylic solvents, and it has been observed that I^- is less nucleophilic than Br^- in acetone solution. Currently, active research programs are aimed at sorting out the way in which the factors of *basicity, polarizability, solvation,* and perhaps other effects all combine in determining the reactivity of a particular nucleophile.

The Nature of the Leaving Group. Since the leaving group is breaking its bond to carbon we would expect that good leaving groups would be weak bases. Strong bases such as OH^- are never the leaving groups in displacement reactions (although H_2O may be a leaving group if a hydroxyl is protonated by acid before displacement). A set of relative average reactivities, compiled by Streitwieser, is shown in Table 3–5.

TABLE 3–5 ▪

Relative Displacement Rates for Leaving Groups

Leaving Group	k_X/k_{Br} (average)
$OSO_2C_6H_5$	6
I	3
Br	1.0
OH_2^+	1
$\overset{+}{S}(CH_3)_2$	0.5
Cl	0.02
ONO_2	0.01
F	0.0001

There is considerable variation in these values, alkyl iodides being from 1.2 to 36 times as reactive as alkyl bromides, depending on the exact reaction studied, but they furnish a good general index to reactivity. Particularly striking is the low reactivity of alkyl fluorides compared with the other halides, apparently because the C—F bond is quite strong; the order of reactivity $I > Br > Cl > F$ parallels the carbon-halogen bond strengths. The low reactivity of alkyl nitrates shows that a poor nucleophile does not necessarily make a good leaving group.

The Nature of the Solvent. Since the S_N2 reaction must involve ionic species as starting materials, products, or both, it must generally be conducted in relatively polar media. The exact effect of the polarity of the solvent on the reaction rate depends on the charges on starting materials and transition states, however. Thus if the reaction involves a neutral nucleophile and a neutral substrate, such as the displacement on methyl iodide by trimethylamine, the transition state will be more ionic than the starting materials; consequently the reaction will be faster in more polar solvents.

$$(CH_3)_3N\!: + CH_3{-}I \longrightarrow (CH_3)_3\overset{\delta\oplus}{N}{-}{-}{-}CH_3{-}{-}{-}\overset{\delta\ominus}{I} \longrightarrow (CH_3)_4\overset{\oplus}{N}\quad I^{\ominus}$$

An S_N2 reaction favored by polar solvents

On the other hand, displacement on trimethylsulfonium ion by ethoxide ion is favored by less polar solvents; since the starting material is more ionic than the transition state, polar solvents lower its energy more.

$$CH_3CH_2O^{\ominus} + CH_3{-}\overset{\oplus}{S}(CH_3)_2 \longrightarrow CH_3CH_2\overset{\delta\ominus}{O}{-}{-}{-}{-}CH_3{-}{-}{-}{-}\overset{\delta\oplus}{S}(CH_3)_2 \longrightarrow$$

$$CH_3CH_2OCH_3 + CH_3SCH_3$$

An S_N2 reaction favored by less-polar solvents

Displacement on a neutral molecule R—X by an anion Y^- involves no net change in charge, so the effect of solvent polarity is small. However, the transition state has charge dispersed over several atoms while the starting materials have charge localized. Polar solvents stabilize localized charges more effectively, so in many cases such displacement reactions are somewhat slower in more polar solvents.

S_N1 reactions

The Structure of the Alkyl Group. S_N1 reactions only occur when the intermediate carbonium ion is stabilized. This generally means that the positive charge is distributed over several atoms rather than being concentrated on one. The distribution of charge can occur (1) by *resonance,* as in the allyl cation; (2) by an *inductive effect* in which single bond electrons shift toward the positive carbon, as in *t*-butyl cation (*hyperconjugation* is often invoked for this case as well, but the relative importance of this effect and of the inductive effect is still a matter of controversy); and (3) by electron sharing from a neighboring group, as in the cation derived from mustard gas.

The rate of ionization to form a carbonium ion depends on the relative energies of starting material and *transition state,* of course, so the stability of the product carbonium ion is really not relevant. However, it seems likely that the transition state will closely resemble the carbonium ion. The transition state has a structure intermediate between that of starting material and that of product; it furthermore is the highest energy point along the reaction coordinate. As Hammond has pointed out, it thus seems reasonable that the very unstable transition state should look very much

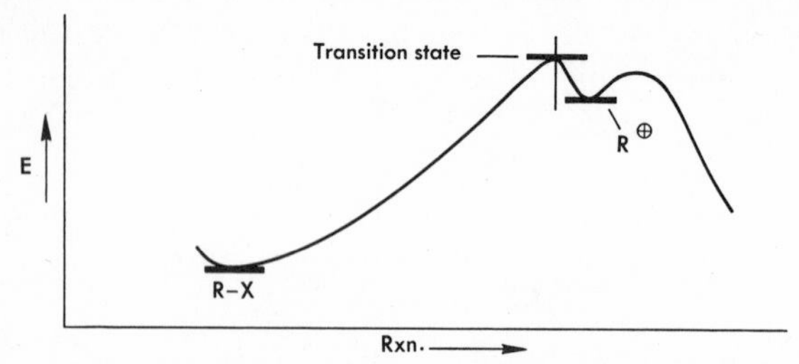

FIGURE 3–1

Energy diagram for the formation of a carbonium ion, in which the transition state strongly resembles the ion.

like the unstable carbonium ion, rather than the stable starting material, so the energy diagram for ionization is that shown in Figure 3–1. For this reason it is common to discuss reaction rates as if the carbonium ion *were* the transition state, although this is not strictly true.

Some indication of the importance of structural factors can be obtained from relative solvolysis rates. In Table 3–6 are listed the relative rates of solvolysis of a series of alkyl bromides in water at 50°C. The precise magnitude of the differences between them depends to some extent on the nature of the solvent (*vide infra*).

Such a table underestimates the effect of methyl groups in facilitating S_N1 reactions, as we go to the more substituted carbonium ions. The solvolysis of the primary bromides probably goes by an S_N2 displacement by the solvent; consequently the S_N1 rate for methyl bromide is much smaller than that shown in the table.

Since benzyl chloride solvolyzes about as readily as does isopropyl chloride, the conjugative effect of one phenyl group stabilizes

TABLE 3–6 ▪

Relative Solvolysis Rates of Alkyl Bromides, in H_2O, 50°C

Compound	Relative Rate
Methyl bromide	1.05
Ethyl bromide	1.00
Isopropyl bromide	11.6
t-Butyl bromide	1,200,000

the transition state about as much as the inductive (or hyperconjugative) effect of two methyls. Of course, as successive phenyls are added the rates increase.

$C_6H_5CH_2$—Cl	$(C_6H_5)_2CH$—Cl	$(C_6H_5)_3C$—Cl
1.0	2000	30,000,000

Relative solvolysis rates in 40% ethanol, 60% ether solution

If a benzyl cation is substituted with electron-donating groups its stability may increase. For instance, *p*-methoxybenzyl chloride solvolyzes at 10,000 times the rate of benzyl chloride in 67% aqueous acetone, but *m*-methoxybenzyl chloride has only ⅔ the rate of benzyl chloride. This is because the positive charge can be stabilized by the *p*-methoxy group but not the *m*-methoxyl. The latter is destabilizing, probably because of a field effect from the carbon-oxygen dipole.

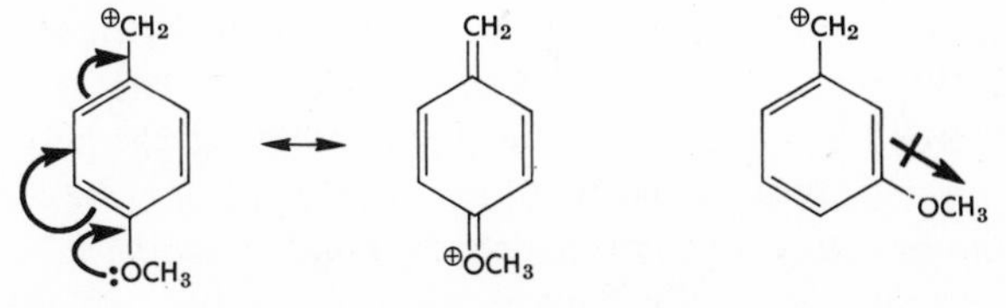

$^{\oplus}CH_2$ … :OCH_3 ⟷ CH_2 … $^{\oplus}OCH_3$ $^{\oplus}CH_2$ … OCH_3

Since the carbon atom of a carbonium ion should be sp^2 hybridized (electrons will enter orbitals with as much *s* character as possible because the *s* orbital is of lower energy) factors which affect the geometry at that carbon can affect S_N1 reactivity. Thus both cyclopropyl and cyclobutyl halides solvolyze slowly, because of the I-strain effect which was encountered in the S_N2 reaction.

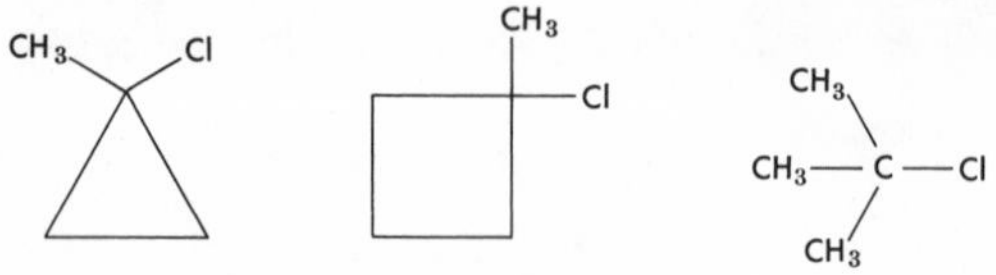

CH_3 Cl CH_3 Cl CH_3 CH_3—C—Cl CH_3

Very slow 0.62 8.9

Relative rates of solvolysis in 80% EtOH

Furthermore, halogens at "bridgeheads" of bridged ring systems ionize very slowly. In a compound such as 1-bromobicycloheptane the resulting carbonium ion cannot flatten, so the carbonium ion is quite unstable. This bromide is also inert in the S_N2 reaction; backside displacement is not possible.

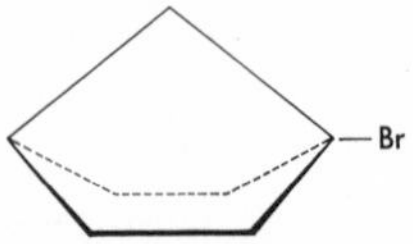

—Br

The Effect of Solvent. Ionization of a neutral compound to an ion pair is strongly favored by polar solvents. The importance of solvent polarity seems to depend on the exact S_N1 reaction being considered, but the effect of solvent on the solvolysis rate of *t*-butyl chloride gives an idea of the magnitudes involved. Solvolysis of *t*-butyl chloride in water at 25°C is 300,000 times faster than solvolysis in ethanol, while mixtures of the two solvents give intermediate rates. In formic acid the rate is 4% of that in water, while in acetone (containing traces of water) hydrolysis is even slower than the reaction in ethanol. The dielectric constant of the solvent is important in determing this reactivity order, but it also depends on the ability of the solvent to interact specifically with the two ions formed. For instance, the anion can be stabilized by hydrogen bonding with a solvent hydroxyl group, while the carbonium ion can be stabilized by specific interaction with an electron pair of a solvent molecule coordinated with it.

$$\mathrm{R{-}Cl} \xrightarrow{\mathrm{H_2O}} \mathrm{H_2O}\text{-----}\overset{\oplus}{\mathrm{R}}\text{-----}\mathrm{OH_2} + \overset{\ominus}{\mathrm{Cl}}\text{---}\mathrm{H{-}O{-}H}$$

Major accelerating effects are also observed when the medium contains a species such as silver ion which can coordinate strongly with a leaving halide ion.

The Nature of the Nucleophile. Since in the S_N1 reaction the nucleophile attacks after the rate-determining step, it cannot affect the rate of reaction. However, if the intermediate carbonium ion has a choice of several nucleophiles, then the nature of the product is determined by their relative nucleophilicities. The same general order of reactivity is found here as was discussed for the S_N2 reaction. Furthermore, many carbonium ions have the possibility of eliminating a proton to form an olefin, and the nucleophilicity of the medium will determine the relative amounts of elimination and substitution. This point will be discussed further in Chapter 4.

General References

A. Streitwieser, Jr., *Solvolytic Displacement Reactions* (McGraw-Hill Book Co., New York, 1962). See also *Chemical Reviews,* **56,** 571 (1956). In spite of the title this is a good general discussion of nucleophilic aliphatic substitution reactions.

J. Hine, *Physical Organic Chemistry* (2nd ed., McGraw-Hill Book Co., New York, 1962), Chapters 6 and 7. Mechanisms and rates of S_N reactions are extensively and critically treated.

C. A. Bunton, *Nucleophilic Substitution at a Saturated Carbon Atom* (Elsevier Publishing Co., New York, 1963). One of a series of short monographs; contains many recent references.

E. Eliel, "Substitution at Saturated Carbon Atoms," in M. Newman, ed., *Steric Effects in Organic Chemistry* (John Wiley & Sons, New York, 1956). A survey of nucleophilic, electrophilic, and free radical mechanisms with emphasis on stereochemistry.

C. K. Ingold, *Structure and Mechanism in Organic Chemistry* (Cornell University Press, Ithaca, New York, 1953), Chapter 7. A good account of the early work in this field.

R. Dewolfe and W. Young, "Substitution and Rearrangement Reactions of Allylic Compounds," *Chemical Reviews,* **56,** 753 (1956). A good review, but a little out-of-date.

B. Capon, "Neighboring Group Participation," *Quarterly Reviews,* **18,** 45 (1964). Contains many recent references.

J. Bunnett, "Nucleophilic Reactivity," *Annual Review of Physical Chemistry,* **14,** 271 (1963). A summary of all the factors which determine nucleophilic reactivity, and the evidence on their importance.

A. Parker, "The Effects of Solvation on the Properties of Anions in Dipolar Aprotic Solvents," *Quarterly Reviews,* **16,** 163 (1962). Includes material on relative nucleophilicities when hydrogen-bonding effects are absent.

F. Jensen and B. Rickborn, *Electrophilic Substitution of Organomercurials* (McGraw-Hill, New York, 1968). An account of an area we have neglected in this book for lack of space.

3

Special Topic

▪ CARBONIUM ION REARRANGEMENTS[1]

IN THE PRECEDING chapter we have furnished examples of neighboring group participation by oxygen, nitrogen, sulfur, etc. There is currently much interest in carbon as a neighboring participating atom; this interest is associated with attempts to understand the intimate mechanisms of carbonium ion rearrangements. We shall examine the evidence that alkyl, phenyl, and vinyl groups can, by migration, directly assist in the ionization step of an S_N1 reaction. First, however, we must survey the types of carbonium ion rearrangements in which such neighboring group assistance can be postulated.

Carbonium ion intermediates, as in S_N1 substitutions (and also E1 eliminations and electrophilic additions to double bonds, Chap-

1. This topic is reviewed by (a) Y. Pocker, "Wagner-Meerwein and Pinacolic Rearrangements in Acyclic and Cyclic Systems," (b) J. Berson, "Carbonium Ion Rearrangements in Bridged Bicyclic Systems," (c) R. Breslow, "Rearrangements in Small Ring Compounds," and (d) J. King and P. de Mayo, "Terpenoid Rearrangements," all in P. de Mayo, ed., *Molecular Rearrangements* (Interscience Publishers, New York, 1963). Other chapters also contain relevant material.

ter 4), frequently undergo rearrangements. For example, in the nitrous acid deamination of neopentylamine,[2] a methyl group migrates with its electron pair so as to form a rearranged more stable carbonium ion. As we shall discuss further below, it is not really clear that the primary carbonium ion is formed first, since rearrangement may be simultaneous with loss of nitrogen.

$$CH_3-\overset{\overset{\displaystyle CH_3}{|}}{\underset{\underset{\displaystyle CH_3}{|}}{C}}-CH_2-NH_2 \xrightarrow{HNO_2} CH_3-\overset{\overset{\displaystyle CH_3}{|}}{\underset{\underset{\displaystyle CH_3}{|}}{C}}-CH_2\overset{\oplus}{N_2} \longrightarrow CH_3-\overset{\overset{\displaystyle CH_3}{|}}{\underset{\underset{\displaystyle CH_3}{|}}{C}}-\overset{\oplus}{CH_2} + N_2 \longrightarrow$$

$$\begin{matrix}CH_3 \\ \diagdown \\ \diagup \\ CH_3\end{matrix}\!\!\overset{\oplus}{C}-CH_2\diagup CH_3 \longrightarrow \begin{matrix}CH_3 \\ \diagdown \\ \diagup \\ CH_3\end{matrix}\!\!\underset{\underset{\displaystyle OH}{|}}{C}-CH_2CH_3 + \begin{matrix}CH_3 \\ \diagdown \\ \diagup \\ CH_3\end{matrix}\!\!C=CH\diagup CH_3$$

Many examples of such alkyl migrations are known, another familiar case[3] being the conversion of pinacol to pinacolone.

$$CH_3-\overset{\overset{\displaystyle CH_3}{|}}{\underset{\underset{\displaystyle OH}{|}}{C}}-\overset{\overset{\displaystyle CH_3}{|}}{\underset{\underset{\displaystyle OH}{|}}{C}}-CH_3 \xrightarrow{H^+} CH_3-\overset{\overset{\displaystyle CH_3}{|}}{\underset{\underset{\displaystyle OH}{|}}{C}}-\overset{\overset{\displaystyle CH_3}{|}}{\underset{\oplus}{C}}-CH_3 \longrightarrow$$

$$CH_3-\overset{\oplus}{\underset{\underset{\displaystyle OH}{|}}{C}}-\overset{\overset{\displaystyle CH_3}{|}}{\underset{\underset{\displaystyle CH_3}{|}}{C}}-CH_3 \longrightarrow CH_3-\underset{\underset{\displaystyle O}{\|}}{C}-\overset{\overset{\displaystyle CH_3}{|}}{\underset{\underset{\displaystyle CH_3}{|}}{C}}-CH_3$$

The alkyl group which migrates may also be part of a ring, in which case ring expansions and contractions result.[4]

2. M. Freund and F. Lenze, "Ein Versuch zur Darstellung des letzten unbekannten Amylalkohols," *Chemische Berichte,* **24,** 2150 (1891).
3. Cf. (a) J. Hine, *Physical Organic Chemistry* (2nd ed., McGraw-Hill Book Co., New York, 1962), Chapter 14; (b) C. Ingold, *Structure and Mechanism in Organic Chemistry* (Cornell University Press, Ithaca, New York, 1953), Chapter 9.
4. Cf. ref. 1c, and also C. Gutsche and D. Redmore, *Carbocyclic Ring Expansion Reactions* (Academic Press, New York, 1968).

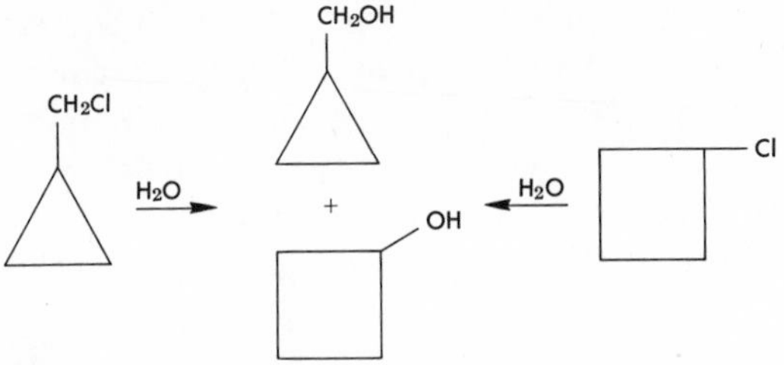

Thus solvolysis of cyclopropylcarbinyl chloride affords a mixture of the unrearranged alcohol and the cyclobutanol from ring expansion of the intermediate carbonium ion, while cyclobutyl chloride yields the same mixture, the cyclopropyl carbinol coming from ring contraction. Some ring opening also occurs.

Many other cases of ring expansions and contractions were discovered in the course of investigating the chemistry of terpenes.[5] As an example, camphene hydrochloride (I) is equilibrated with isobornyl chloride (II) on treatment with Lewis acids, which can help remove chloride ion and promote carbonium ion formation.

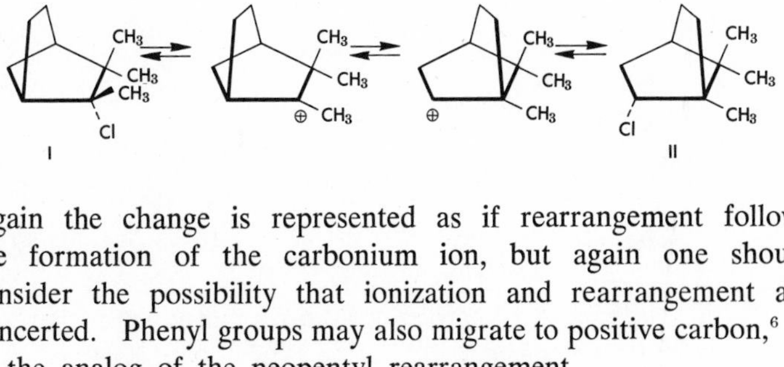

Again the change is represented as if rearrangement follows the formation of the carbonium ion, but again one should consider the possibility that ionization and rearrangement are concerted. Phenyl groups may also migrate to positive carbon,[6] as in the analog of the neopentyl rearrangement.

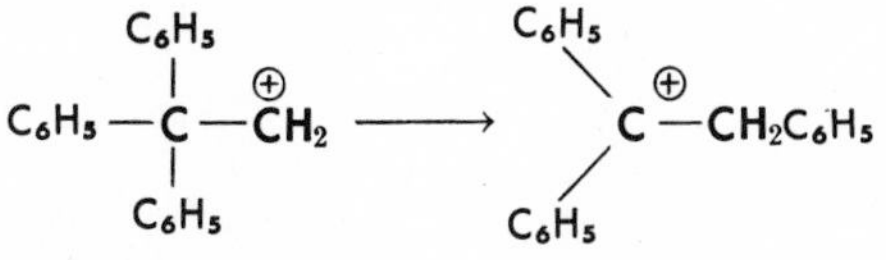

$$(C_6H_5)_3C-\overset{\oplus}{C}H_2 \longrightarrow (C_6H_5)_2\overset{\oplus}{C}-CH_2C_6H_5$$

5. Cf. refs. 1d and 1b.
6. Cf. D. Cram, "Intramolecular Rearrangements," in M. Newman, ed., *Steric Effects in Organic Chemistry* (John Wiley & Sons, New York, 1956), Chapter 5.

Carbonium ions can undergo ring closure with double bonds. A simple example[7] occurs in the hydrolysis of 5-chloro-2-methyl-2-pentene (V), affording almost entirely cyclopropyldimethyl carbinol. Other examples are the acetolysis of *exo*-norbornenyl "brosylate" (*p*-bromobenzenesulfonate) to afford chiefly the cyclized product (VI),[8] and the methanolysis of cholesteryl chloride (VII) to yield i-cholesteryl methyl ether.[9]

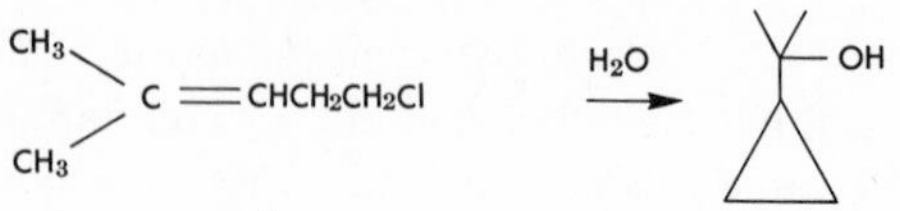

$$(CH_3)_2C{=}CHCH_2CH_2Cl \ (V) \xrightarrow{H_2O} \text{cyclopropyl}{-}C(CH_3)_2OH$$

V

AcOH AcO—

OSO₂ Br

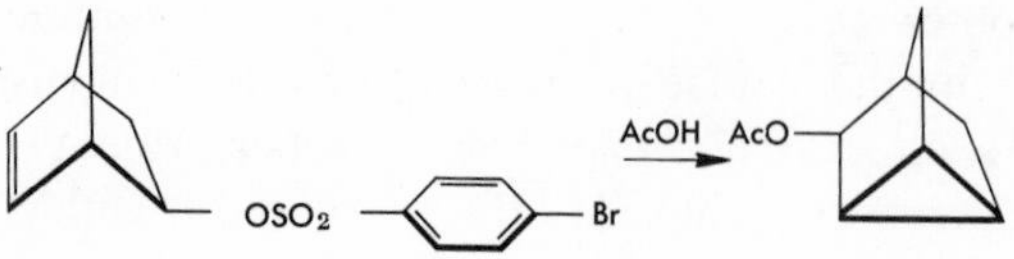

VI

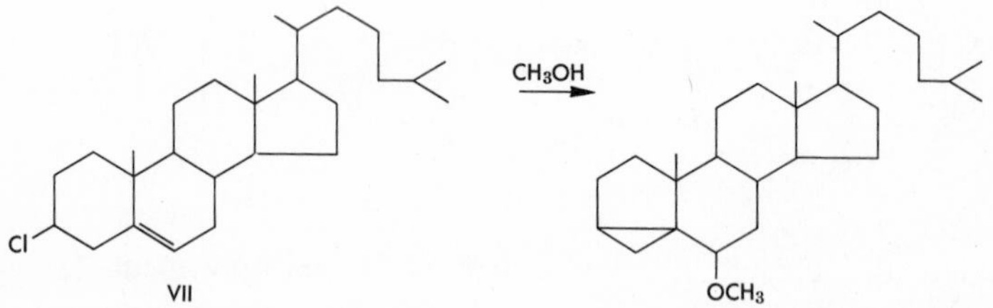

CH₃OH

Cl

VII

OCH₃

Among other types of carbonium ion rearrangements, there are many examples of hydride shifts to neighboring carbonium ions.[10] A rather unusual one[11] occurs in the reaction of cycloöctene oxide

7. Ref. 1c, p. 260.
8. Ref. 3a, p. 323; ref. 1b, p. 192.
9. N. Wendler, "Rearrangements in Steroids," in P. de Mayo, ed., *Molecular Rearrangements* (Interscience Publishers, New York, 1964), Chapter 16, p. 1075.
10. Ref. 3a, p. 330.
11. A. Cope, S. Fenton, and C. Spencer, "Molecular Rearrangement of Cycloöctene Oxide on Solvolysis," *Journal of the American Chemical Society,* **74,** 5884 (1952).

with formic acid, followed by hydrolysis. A mixture is obtained of the expected *trans*-1,2-cyclooöctanediol and of *cis*-1,4-cyclooöctanediol. The latter compound evidently arises from a transannular shift of hydride (hydrogen with its electron pair) to the carbonium ion first formed.

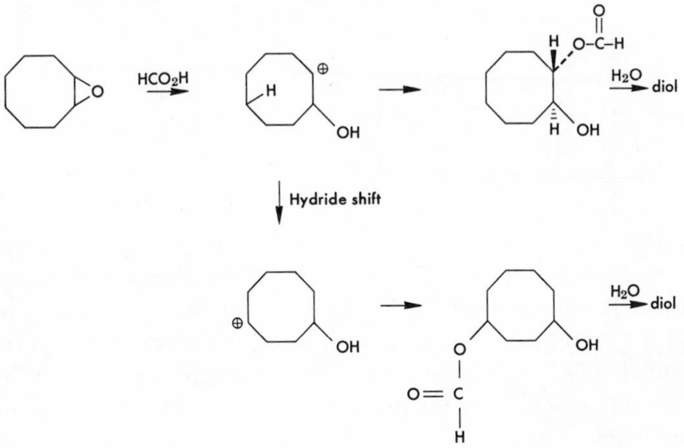

HCO_2H → ⊕ H OH → H O–C(=O)–H, H OH → H_2O → diol

↓ Hydride shift

⊕ OH → O–C(=O)–H, OH → H_2O → diol

Although the distance seems large in this drawing, models show that the migrating hydrogen can actually be quite close to the positive carbon.

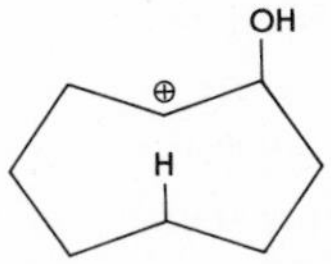

OH ⊕ H

Rearrangements are also common in which not a carbonium ion but a related species is involved. Thus the Beckmann rearrangement[12,13] of oximes involves migration to a positive nitrogen, rather than carbon.

12. L. G. Donaruma and W. Heldt, "The Beckmann Rearrangement," in A. C. Cope, ed., *Organic Reactions,* Vol. 11 (John Wiley & Sons, New York, 1960), p. 1.
13. Cf. P. Smith, "Rearrangements Involving Migration to an Electron-deficient Nitrogen or Oxygen," in P. de Mayo, ed., *Molecular Rearrangements* (Interscience Publishers, New York, 1963), Chapter 8.

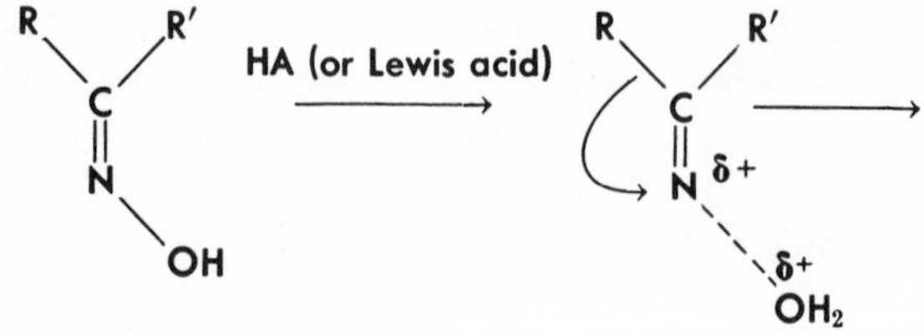

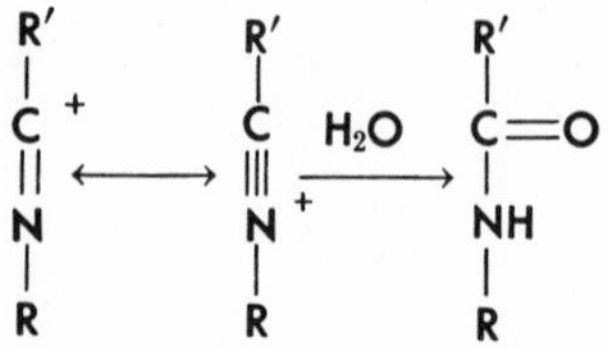

Migration to positive oxygen is also known, as in the rearrangement of *trans*-9-decalyl perbenzoate.[13] The benzoate ion remains intimately associated with the cation at all times; the externally added benzoate anion fails to compete—this failure to incorporate external anions is the usual evidence for internal return in an ion pair—and interestingly the two oxygens of the benzoate ion do not even equilibrate.

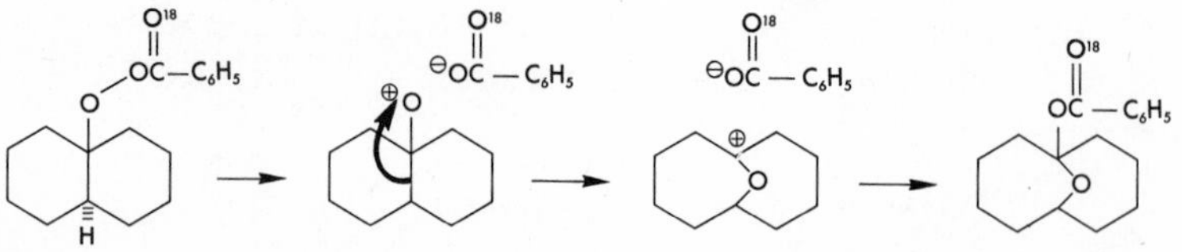

For clarity the reaction is written as if the unrearranged ion were first formed, although it seems very likely that rearrangement accompanies ionization. Finally, rearrangements involving carbenes and nitrenes are related to these processes. Examples are the Wolff rearrangement[14] of diazoketones (A); the similar Curtius degradation[14,15] via acyl azides (B); and the Hofmann degradation[14,16] (C).

14. Ref. 13, p. 528.
15. P. Smith, "The Curtius Reaction," in R. Adams, ed., *Organic Reactions,* Vol. 3 (John Wiley & Sons, New York, 1946), p. 337.
16. E. Wallis and J. Lane, "The Hofmann Reaction," in R. Adams, ed., *Organic Reactions,* Vol. 3 (John Wiley & Sons, New York, 1946), p. 267.

$$\text{(A)}\quad R{-}\overset{\overset{\displaystyle O}{\|}}{C}{-}\overset{\ominus}{C}H{-}\overset{\oplus}{N}{\equiv}N \xrightarrow{\Delta} R{-}\overset{\overset{\displaystyle O}{\|}}{C}{-}\ddot{C}H \longrightarrow R{-}CH{=}C{=}O \xrightarrow{H_2O} RCH_2CO_2H$$

$$\text{(B)}\quad R{-}\overset{\overset{\displaystyle O}{\|}}{C}{-}\overset{\ominus}{N}{-}\overset{\oplus}{N}{\equiv}N \xrightarrow{\Delta} R{-}\overset{\overset{\displaystyle O}{\|}}{C}{-}\ddot{N} \longrightarrow R{-}N{=}C{=}O$$

$$\text{(C)}\quad R{-}\overset{\overset{\displaystyle O}{\|}}{C}{-}\underset{\displaystyle Br}{NH} \xrightarrow{OH^-} R{-}\overset{\overset{\displaystyle O}{\|}}{C}{-}\underset{\displaystyle Br}{\overset{\ominus}{N}} \longrightarrow R{-}\overset{\overset{\displaystyle O}{\|}}{C}{-}\ddot{N} \longrightarrow R{-}N{=}C{=}O$$

Anchimeric Assistance

We have written most of these rearrangements as if ionization occurred first. However, in many cases it is known that the migration actually assists ionization; both rates and stereochemistry support this idea. This process, neighboring group assistance by a carbon atom, has been called[17] "anchimeric assistance." Thus in the Beckmann rearrangement it is clear that an unrearranged nitrogen cation is not formed first since either group should then migrate with equal likelihood; the experimental result is that the group migrates which is *trans* to the leaving group.[12]

$$\begin{matrix} R \quad R' \\ \diagdown\;\diagup \\ C \\ \| \\ N \\ \diagdown \\ \overset{\oplus}{O}H_2 \end{matrix} \longrightarrow \begin{matrix} R' \\ | \\ C^{\oplus} \\ \| \\ N \\ \diagup \\ R \end{matrix} \quad \text{Directly}$$

If the nitrogen cation were strongly associated with the leaving group, as a tight ion pair, there might still be selective stereochemistry. However, this seems very unlikely because the reaction occurs under conditions which are undoubtedly too mild to generate an unstabilized nitrogen cation. With simultaneous mi-

17. S. Winstein, C. Lindegren, H. Marshall, and L. Ingraham, "Participation in Solvolysis of Some Primary Benzenesulfonates," *Journal of the American Chemical Society*, **75**, 147 (1953).

gration a much better cation is produced; the ionization occurs with neighboring group participation by the phenyl.

As another example, β,β,β-triphenylethyl chloride solvolyzes in formic acid at a rate 60,000 times that of neopentyl chloride.[18] Phenyls withdraw electrons inductively, so they would destabilize the already unstable primary cation or a transition state which resembled the unrearranged cation, but if migration is simultaneous with ionization the rate can be understood.

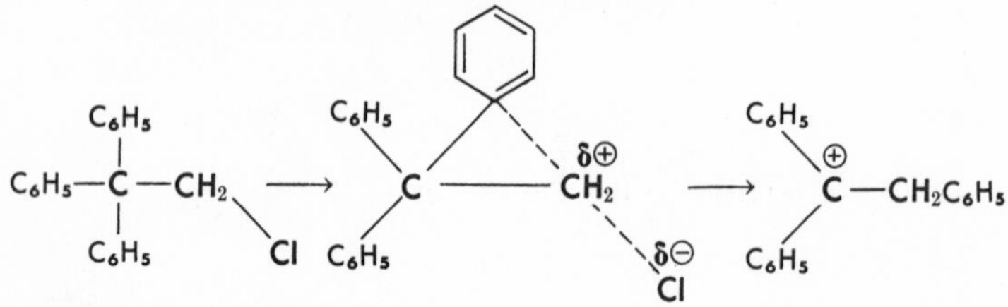

Open versus Bridged Ions[19]

In the two cases just considered it was apparent that the transition state for ionization could not simply resemble the unrearranged carbonium ion or nitrogen cation. Instead rearrangement occurred simultaneously with ionization, and it was suggested that the actual product was directly the rearranged carbonium ion

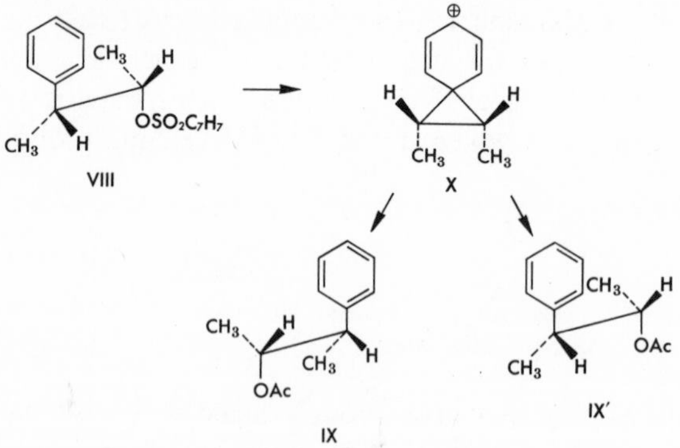

18. Ref. 3b, p. 514; cf. S. Winstein, B. Morse, F. Grunwald, K. Schreiber, and J. Corse," "Driving Forces in the Wagner-Meerwein Rearrangement," *Journal of the American Chemical Society,* **74,** 1113 (1952).
19. For a discussion of the history of this question with a collection of key reprints see P. Bartlett, *Nonclassical Ions* (W. A. Benjamin, Inc., New York, 1965).

(nitrogen cation). However, some even more unusual situations are known. For instance, the acetolysis (solvolysis in acetic acid) of optically active threo-2-phenyl-3-butyl toluenesulfonate (VIII) affords the racemic threo acetate (IX and IX′).[6]

The formation of only the threo acetate shows that the toluenesulfonate group is replaced with *retention* of configuration, as in other examples of neighboring group participation. The formation of completely racemized product is expected from the ion X, which has a plane of symmetry. An ion of this kind has been called[20] a "phenonium" ion. Further evidence that X is the direct product of ionization comes from the finding[21] that 2-*p*-anisyl-3-butyl toluenesulfonate (XI) is 80 times as reactive as is the phenyl compound. Thus the transition state for ionization at least partially resembles the phenonium ion.

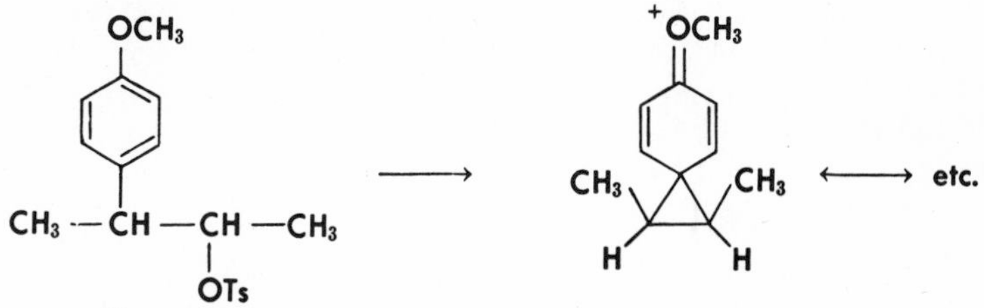

OCH_3

$CH_3—CH—CH—CH_3$

OTs

XI

$\longrightarrow$

$^+OCH_3$

CH_3 CH_3

H H

$\longleftrightarrow$ etc.

Of course, anchimeric assistance by a phenyl ring must compete with simple nucleophilic displacement by solvent. Thus if the solvent is very nucleophilic, phenyl participation may not compete well,[22] while in a poorly nucleophilic solvent such as trifluoroacetic acid the path involving phenyl participation may be much faster.[23]

20. D. Cram, "Phenonium Sulfonate Ion-pairs. . . .", *Journal of the American Chemical Society,* **74,** 2129 (1952).
21. S. Winstein, M. Brown, K. Schreiber, and A. Schlesinger, "Neighboring Carbon and Hydrogen. IX," *Journal of the American Chemical Society,* **74,** 1140 (1952).
22. For a discussion, with many references, of the case against phenonium ions see H. C. Brown and C. Kim, "Structural Effects in Solvolytic Reactions. III.—" *Journal of the American Chemical Society,* **90,** 2082 (1968).
23. J. Nordlander and W. Deadman, "Trifluoroacetolysis of 2-Phenylethyl *p*-Toluenesulfonate. Evidence for Phenonium Ion," *Tetrahedron Letters,* 4409 (1967).

The bridged species has even been trapped in a special case, solvolysis of XII to afford XIII.[24] Treatment of XIII with

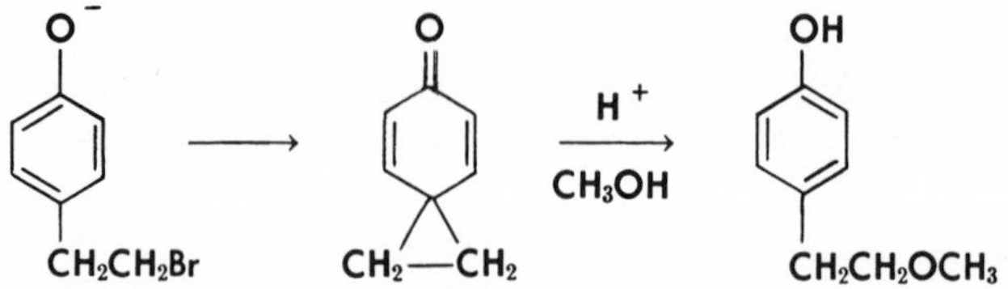

$$\underset{\text{XII}}{p\text{-}O^{-}C_6H_4CH_2CH_2Br} \longrightarrow \underset{\text{XIII}}{\text{spiro}[2,5]\text{octa-1,4-diene-3-one } (CH_2\text{—}CH_2)} \xrightarrow[CH_3OH]{H^+} p\text{-}HOC_6H_4CH_2CH_2OCH_3$$

acidic methanol converts it to the normal open-chain product. Such phenonium ions are presumably part of the reaction path in any carbonium ion migration of a phenyl group, but it is interesting that in solvolysis of XI the bridged ion is more stable than either open carbonium ion. This follows from the fast solvolysis of XI, which shows that bridging stabilizes the ion. However, strictly speaking it is not proved that a *symmetrical* ion is

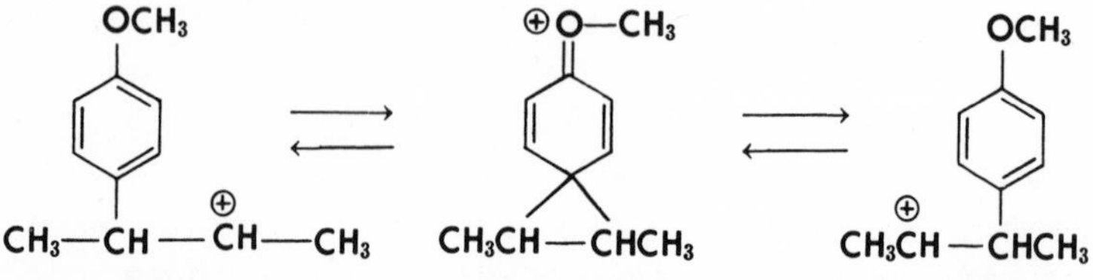

$$\underset{\text{Less stable}}{CH_3\text{—}CH(C_6H_4OCH_3)\text{—}\overset{\oplus}{C}H\text{—}CH_3} \rightleftharpoons \underset{\text{More stable}}{\text{bridged } \oplus O\text{—}CH_3 \text{ dienone, } CH_3CH\text{—}CHCH_3} \rightleftharpoons \underset{\text{Less stable}}{CH_3\overset{\oplus}{C}H\text{—}CH(C_6H_4OCH_3)CH_3}$$

the best; the lowest energy species could be an *unsymmetrical* ion, with the anisyl group strongly bonded to one carbon and only weakly bonded to the other.

When alkyl groups migrate in carbonium ion rearrangements, an intermediate bridged ion can also be written, often called a "nonclassical" carbonium ion. The structure of this intermediate has been written with dotted bonds to carbon; it can also be represented in molecular orbital terms. The combination of a (hybrid) atomic orbital from the methyl carbon with an atomic orbital from

24. R. Baird and S. Winstein, "Isolation and Behavior of *spiro*[2,5] Octa-1,4-diene-3-one," *Journal of the American Chemical Society*, **79,** 4238 (1957).

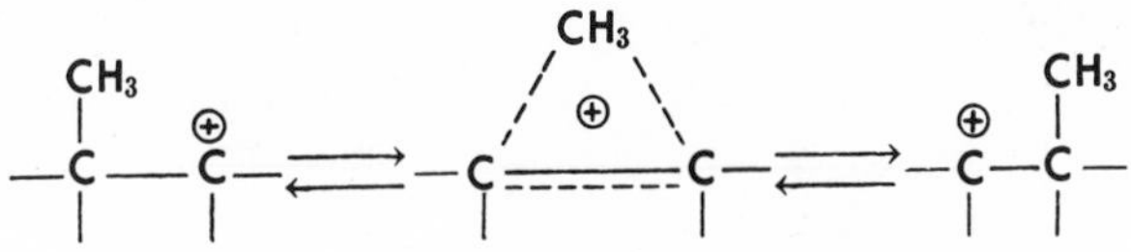

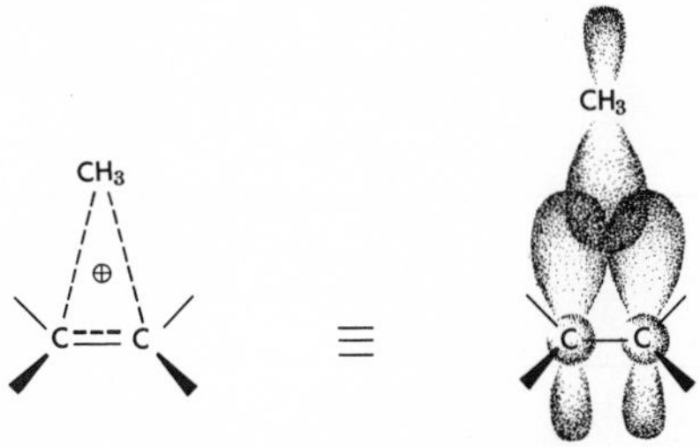

the other two carbons leads to a new molecular orbital which can accommodate the two electrons (of the erstwhile carbon-methyl bond). The lack of similar alkyl migrations in radicals or in carbanions is explained by the fact that this m.o. has room for only two electrons, not the three or four which would be present in a similar intermediate in radical or anion rearrangements.[25] Much recent research and discussion[19,26] has centered on the question of whether such a bridged carbonium ion can be more stable than either simple ion to which it is related.

The question is best illustrated further by examining a particular case, the solvolysis of 2,2,1-bicycloheptyl bromobenzenesulfonate (XIV) in acetic acid at 25°C.[27] Although the starting material is optically active, migration of the two carbon bridge causes racemization (since structure XVI halfway through the migration has a plane of symmetry, and interconverts XV and XV′ which are mirror images).

25. H. Zimmerman and A. Zweig, "Carbanion Rearrangements," *Journal of the American Chemical Society,* **83,** 1196 (1961).
26. See, for instance, H. C. Brown, "The Norbornyl Cation—Classical or Non-classical," *Chemistry in Britain,* 199 (1966).
27. Ref. 1b, p. 123, and references therein; cf. also G. D. Sargent, "Bridged, Non-classical Carbonium Ions," *Quarterly Reviews,* **20,** 301 (1966).

H, O, SO_2, Br — XIV ⟶ XV ⇌ XVI ($\frac{1}{2}\oplus$, $\frac{1}{2}\oplus$) ⇌ XV′ ⟶ XVII (OAc) + XVII′ (OAc)

The observed result is that completely racemic acetate is produced in this solvolysis, i.e., an equal mixture of XVII and XVII′. One explanation of this result would be the scheme shown, in which a simple carbonium ion XV is first formed, this then rearranges to the ion XV′ (its mirror image), and the resulting racemic mixture of carbonium ions then reacts with acetic acid exclusively on the unhindered bottom of the ring to yield the racemic mixture of products. The other possibility is that the bridged ion XVI is more stable than the open ion XV, and that solvolysis leads directly to XVI. Reaction of XVI with acetic acid would then produce the racemic mixture of acetates, and the classical ions XV and XV′ would never have been involved at all.

XIV (OSO_2, Br) ⟶ XVI ($\frac{1}{2}\oplus$ $\frac{1}{2}\oplus$) $\xrightarrow{\text{HOAc}}$ XVII + XVII′

Other evidence had long been interpreted to indicate that this latter possibility is the correct one, and that the solvolysis does not involve classical ions. For instance, acetolysis of XIV (called the exo isomer) is 350 times as fast as acetolysis of XVIII, its epimer. Both could lead to the same classical ion, XV, but only XIV can

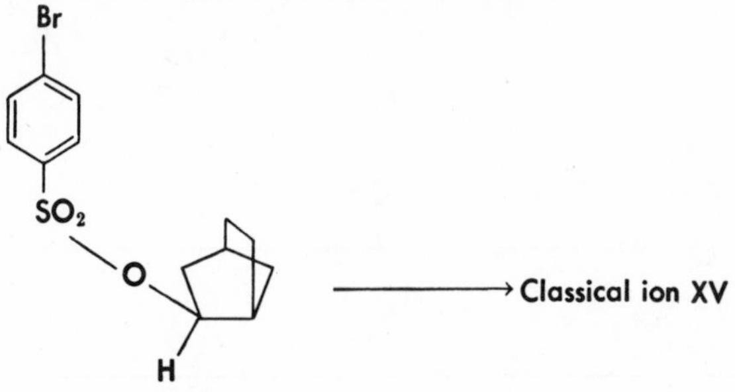

Br
SO_2
O
H
→ Classical ion XV
XVIII

ionize with simultaneous participation of the migrating group at the back of the carbon, leading directly to XVI, a bridged ion. This direct conversion of XIV to the better cation was considered to be the explanation of the large acceleration.

Although formation of the nonclassical ion may contribute to the difference in solvolysis rates of XIV and XVIII, Brown has produced effective arguments[26,28] that there must be major steric effects as well. Recently Schleyer[29] has pointed out a particular kind of steric effect which differs in the transition states for solvolysis of XIV and XVIII and which could account for at least some of the observed rate difference. Accordingly, in spite of a large amount of work on this system, the importance of the nonclassical ion XVI is still far from clear.

On the other hand, it seems quite clear that solvolysis of cyclopropylcarbinyl chloride (XIX) leads to a "nonclassical" carbonium ion.[30] Thus this compound is 40 times as reactive as is 2-methylallyl chloride (XX), although the latter compound can lead to a highly stabilized allylic cation. The cyclopropylcarbinyl cation, if it were a simple saturated primary cation, would be very unstable; the high reactivity of the system can be explained if the product of solvolysis is a bridged ion instead.

28. H. C. Brown and K. Takeuchi, "The Characteristics of a Highly Stabilized, Classical Norbornyl Cation," *Journal of the American Chemical Society,* **90,** 2691 (1968).
29. P. Schleyer, "Torsional Effects in Polycyclic Systems," *Journal of the American Chemical Society,* **89,** 701 (1967).
30. Ref. 1c, p. 259.

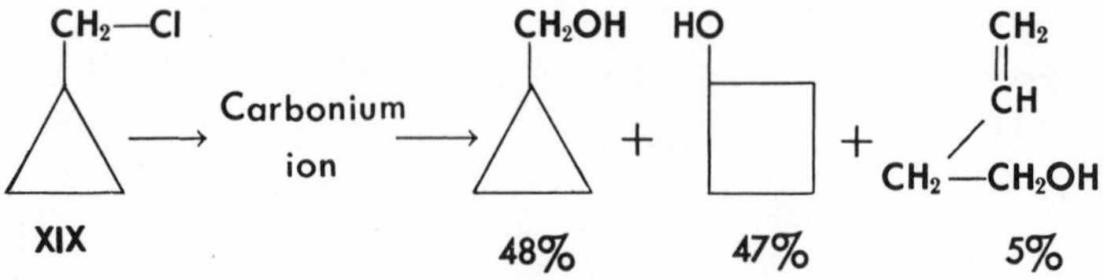

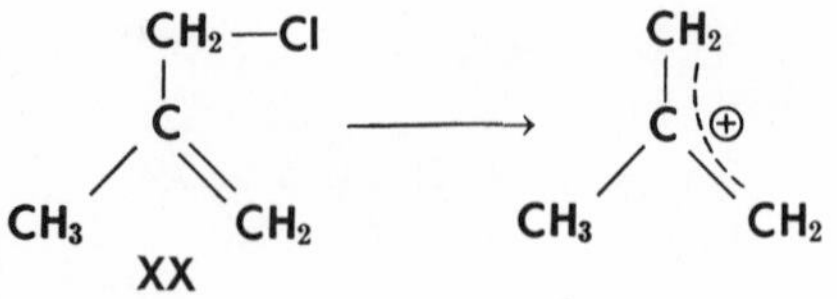

$$\text{XIX (cyclopropyl-}CH_2\text{—Cl)} \xrightarrow{\text{Carbonium ion}} \text{cyclopropyl-}CH_2OH\ (48\%) + \text{HO-cyclobutane}\ (47\%) + CH_2{=}CH{-}CH_2{-}CH_2OH\ (5\%)$$

$$\text{XX: } CH_3{-}C(CH_2{-}Cl){=}CH_2 \longrightarrow CH_3{-}\overset{\oplus}{C}(CH_2)\cdots CH_2$$

Although single-bond electrons are participating in the intermediate ion, they come from the bent bonds of a cyclopropane and are not typical σ electrons. A study[31] of structural and substituent effects on reaction rates shows that the transition state involves overlap of the carbonium ion p orbital with both adjacent C—C single bonds.

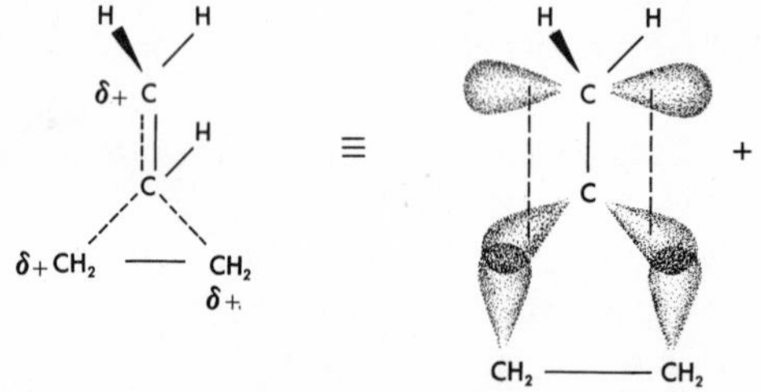

H H δ+ C H C δ+CH₂ CH₂ δ+ ≡ H H C C CH₂ CH₂ +

Here the rearrangement leads to a new structure, so it might be wondered whether one does not simply have ordinary anchimeric assistance by a ring-expanding migration.

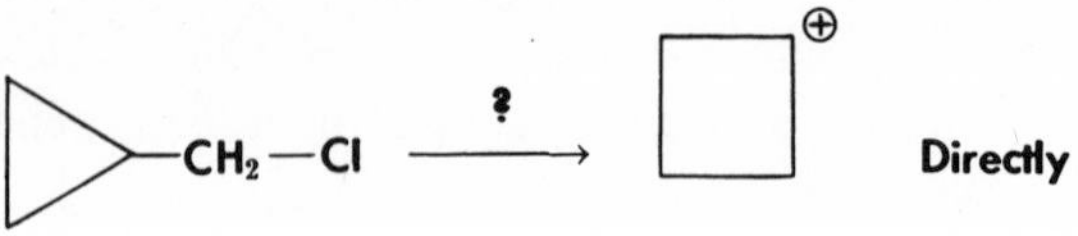

$$\text{cyclopropyl-}CH_2\text{—Cl} \xrightarrow{?} \text{cyclobutyl}^{\oplus} \quad \textbf{Directly}$$

31. P. Schleyer and G. van Dine, "Substituent Effects on Cyclopropylcarbinyl Solvolysis Rates," *Journal of the American Chemical Society,* **88,** 2321 (1966).

Part of the evidence against this is the fact that since almost half the product is unrearranged alcohol, the product of ionization is *not* a classical cyclobutyl cation.

A nonclassical ion is also formed in the solvolysis of cyclobutyl chloride, whose rate is also abnormally fast. Although this ion is apparently not identical with the cyclopropylcarbinyl cation, the two ions rapidly interconvert: Precisely the same product mixture is formed starting from either cyclopropylcarbinyl chloride or cyclobutyl chloride.[30]

Ordinary double bonds can also participate directly in the ionization step in certain compounds. One of the most striking

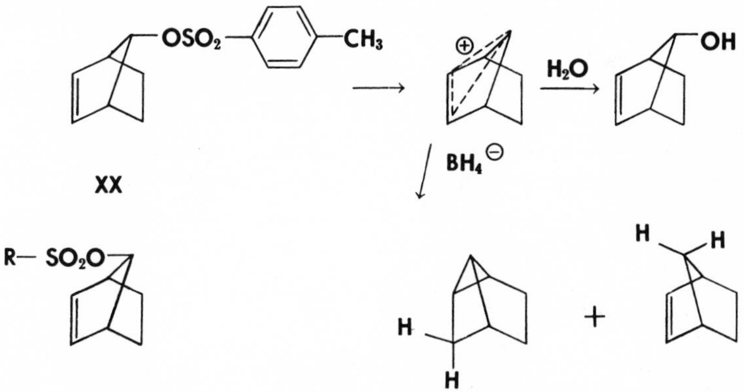

OSO_2—C6H4—CH_3 ⟶ ⊕ ⟶ (H_2O) OH

XX

$BH_4^{\ominus}$

R— SO_2O

XXI

H H H H +

XXII

examples of this is found[32] in the solvolysis of 7-*anti*-norbornenyl toluenesulfonate (XXI). The rate is 10,000,000 times that for the *syn* isomer XXII, showing that the double bond participates in the ionization step. When the product carbonium ion is trapped by reaction with borohydride ion[33] a large amount of the strained hydrocarbon XXIII is formed, together with norbornene (XXIV). With water as the nucleophile only 7-*anti*-norbornenol is formed.[32]

32. Ref. 1b, p. 196.
33. H. Brown and H. Bell, "The Reaction of 7-Norbornadienyl and 7-Dehydronorbornyl Derivatives with Borohydride under Solvolytic Conditions—Evidence for the Tricyclic Nature of the Corresponding Cations," *Journal of the American Chemical Society,* **85,** 2324 (1963).

Although we have written the above carbonium ion as "non-classical," it has been argued,[33] and disputed,[34] that the system involves simply anchimeric assistance by a double bond during ionization.

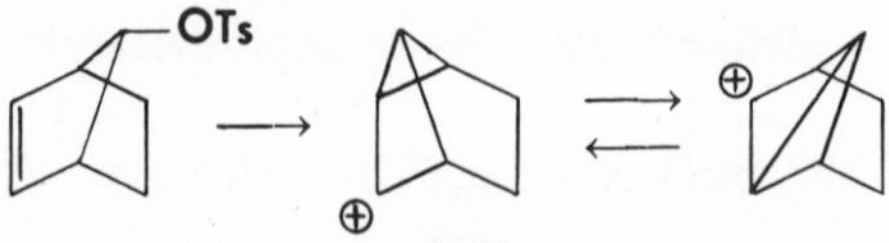

OTs → ⊕ XXV ⇌ ⊕

Since the equilibrating "classical" ions (XXV) are cyclopropylcarbinyl cations, which as we have noted above are themselves "nonclassical," the distinction between the two interpretations is not a major one.

We have emphasized cases in which there is good evidence for neighboring group participation by carbon, but it should be pointed out that such situations are relatively rare. When ionization of a compound leads directly to a stable classical cation, then anchimeric assistance is not observed. Accordingly, nonclassical ions are not of frequent occurrence in organic chemistry. However, the subtlety of some of the questions being asked continues to stimulate research in this area.

34. S. Winstein, A. Lewin, and K. Pande, "The Non-Classical 7-Norbornenyl Cation," *Journal of the American Chemical Society,* **85,** 2324 (1963).

4

IONIC ELIMINATION AND ADDITION REACTIONS

OLEFINS ARE GENERALLY synthesized by elimination reactions from saturated compounds; conversely, the most characteristic reactions of olefins are additions to the double bond. Thus there would seem to be a logical connection between these two classes of reactions. The connection is actually much more fundamental: a reaction and its reverse must occur over the same path, although in opposite directions. Since there is only one lowest energy path between A and B, it is traveled both for A → B and for B → A, as shown in Figure 4–1. This very important rule is called the *principle of microscopic reversibility.* It means that the elucidation of the mechanism of an elimination reaction would simultaneously furnish the mechanism of the reverse, addition, reaction, *provided both occur under the same conditions.* This last is an important limitation. For instance, the addition of HBr to an olefin, under acidic conditions, is not the reverse of the *base-catalyzed* elimination of HBr to form the olefin; in the presence of base a new pathway (*vide infra*) is made lower in energy.

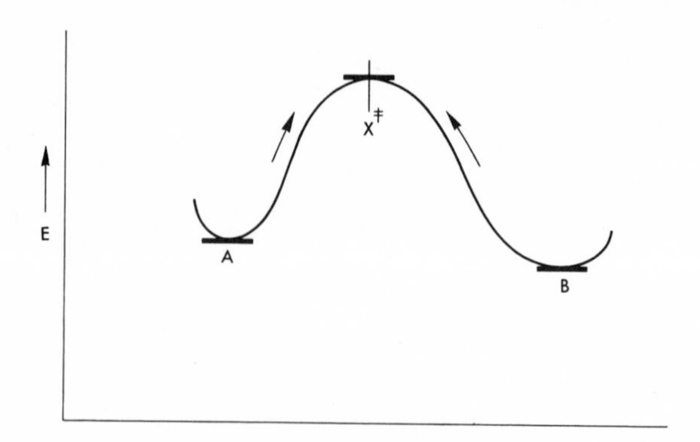

FIGURE 4–1

The common path and common transition state for a forward and reverse reaction run under the same conditions.

However, there are other reactions for which the principle is quite useful; alcohols can be dehydrated with acid, and olefins can be hydrated to alcohols with acid, so that under the proper conditions an equilibrium is established. In this case the mechanism of the hydration step must be exactly the reverse of that of the dehydration.

4–1 Elimination Reactions

In a β-elimination reaction two groups are lost from neighboring atoms, with the resultant formation of a double bond. For convenience we shall symbolize such reactions as if a carbon-

$$-\underset{|}{\overset{\overset{X}{|}}{C}}-\underset{|}{\overset{\overset{Y}{|}}{C}}- \longrightarrow \;\;\gt C=C \lt$$

β-elimination

carbon double bond were formed, although analogous processes may be written involving the formation of carbon-oxygen, carbon-nitrogen, etc., double bonds. Most studies of mechanisms of β-elimination reactions have concentrated on the formation of a carbon-carbon double bond by the elimination of HX, where X is a good leaving group such as a halide ion. Accordingly we shall discuss only such HX eliminations.

Other reactions are known which may be classified as α-eliminations (two groups from the same carbon), and 1,3-eliminations and even 1,4-eliminations are known. Of these, only the α-eliminations, to form carbenes, will be briefly discussed.

$$CCl_3H \xrightarrow{OH^-} Cl_2C: \quad \text{via } CCl_3^-$$

An α-elimination

$$BrCH_2-CH_2-CH_2Br \xrightarrow{Zn} \text{cyclo-}(CH_2)_3 + ZnBr_2$$

A 1,3-elimination

$$H-CH_2-CH=CH-CH_2Br \xrightarrow{HO^-} CH_2=CH-CH=CH_2$$

A 1,4-elimination

β-Eliminations

The E1 Mechanism. If a carbonium ion has a β-hydrogen, this may be lost to yield an olefin. Thus in many of the S_N1 reactions discussed in Chapter 3 olefin formation is an important side reaction. The elimination of HX by ionization of X, and subsequent loss of a proton from the resulting carbonium ion, has been called the E1 mechanism (Elimination Unimolecular). It

$$H-\overset{|}{\underset{|}{C}}-\overset{|}{\underset{|}{C}}-X \xrightarrow{\text{Slow}} H-\overset{|}{\underset{|}{C}}-\overset{|}{\underset{|}{C}}^{\oplus} \xrightarrow[B:]{\text{Fast}} >C=C< + \overset{\oplus}{BH}$$

The E1 mechanism

would be expected that such a mechanism could occur only when the intermediate carbonium ion is stable. Thus the same factors which favor S_N1 reactions—formation of a stabilized carbonium ion in a highly polar medium—will operate here. The E1 elimination and the S_N1 reaction often occur simultaneously; substitution is favored by the presence of good nucleophiles, while strong bases will tend to remove the proton from the carbonium ion and favor elimination.

The evidence for an E1 mechanism is, first of all, kinetic. Since the rate-determining step does not involve the base the reaction rate will be independent of base concentration. However, most E1 eliminations accompany the solvolysis of *t*-alkyl halides in the absence of added base; the solvent molecules must act as proton acceptors, and they cannot be detected kinetically. For this reason a more subtle mechanistic criterion is used. If a reaction follows an E1 mechanism it involves formation of a carbonium ion, and the behavior of this carbonium ion should be independent of the nature of the leaving group. Thus in the competition between S_N1 and E1 reactions a mixture of olefin and alcohol (if the solvent is water) is formed, and the composition of this mixture should be the same whether the starting material is *t*-butyl chloride, *t*-butyl bromide, *t*-butyl iodide, etc., since the same carbonium ion is formed from each one. In general this is found to be the case for reactions of *t*-alkyl halides in quite polar media (although in less polar solvents the leaving group may stay associated with the carbonium ion and exert some influence on the course of further reaction).

$$CH_3-\underset{CH_3}{\overset{CH_3}{C}}-X \xrightarrow[25^\circ]{EtOH} CH_3-\underset{CH_3}{\overset{CH_3}{C}}\oplus \xrightarrow{E1} CH_3-\underset{CH_3}{C}{=}CH_2 \quad (19\%)$$

$$X = Br, Cl, I, \overset{\oplus}{S}Me_2$$

$$CH_3-\underset{CH_3}{\overset{CH_3}{C}}\oplus \xrightarrow{S_N1} CH_3-\underset{CH_3}{\overset{CH_3}{C}}-OEt \quad (81\%)$$

Another criterion which may be applied is the composition of the olefin mixture, in cases where a mixture is possible. Thus solvolysis of *t*-amyl halides yields, in addition to substitution products, a mixture of two olefins, 2-methylbutene-1 and 2-methylbutene-2. The stable more substituted olefin is preferred (so it is said that E1 elimination follows the *Saytzeff rule*), and the observation is made that the proportion of the two olefins is independent of the leaving group in reasonably polar media; again this means that a common intermediate, the carbonium ion, is involved in the eliminations.

$$CH_3-\underset{X}{\overset{CH_3}{C}}-CH_2-CH_3 \xrightarrow[25°]{EtOH} CH_3-\underset{\oplus}{\overset{CH_3}{C}}-CH_2CH_3 \xrightarrow[E1]{37\%}$$

$$\downarrow S_N1 \quad 63\%$$

$$CH_3-\overset{CH_3}{C}=CHCH_3 \; (82\%) + CH_3-\overset{CH_2}{\overset{\|}{C}}CH_2CH_3 \; (18\%)$$

The E2 Mechanism. Most commonly, HX eliminations are carried out in the presence of strong base. For instance, when ethyl bromide is held at 60°C in ethanol nothing happens, since the ethyl cation is not stable enough for E1 elimination to occur. However, in the presence of sodium ethoxide, a strong base, elimination of HBr occurs and ethylene is formed (together with considerable diethyl ether from an S_N2 reaction). Since it is found that the rate of olefin formation is proportional to both the ethyl bromide and the ethoxide ion concentrations, a bimolecular mechanism is written, the E2 mechanism (Elimination Bimolecular).

$$\overset{\ominus}{OEt} \quad H-CH_2-\underset{Br}{CH_2} \longrightarrow CH_2=CH_2 + EtOH + Br^{\ominus}$$

$$\text{Rate} = k\,[C_2H_5Br]\,[OEt^{\ominus}]$$

In this mechanism it is suggested that removal of the proton and loss of the bromide ion occur simultaneously, but of course this is not required by the observed kinetics. The three schemes shown below would also show second-order kinetics, and must be excluded on other grounds. The first one is easily excluded since other experience with carbonium ions suggests that if the ethyl

$$CH_3{-}CH_2{-}Br \underset{}{\overset{\text{Fast}}{\rightleftharpoons}} CH_3{-}\overset{\oplus}{C}H_2 \;\; \overset{\ominus}{Br} \xrightarrow[\text{OEt}^{\ominus}]{\text{Slow}} CH_2{=}CH_2$$

$$\overset{H}{\overset{|}{C}}H_2{-}CH_2{-}Br \underset{\text{Fast}}{\overset{\text{OEt}^{\ominus}}{\rightleftharpoons}} \overset{\ominus}{C}H_2{-}CH_2{-}Br \xrightarrow{\text{Slow}} CH_2{=}CH_2$$

$$\overset{H}{\overset{|}{C}}H_2{-}CH_2{-}Br \xrightarrow[\text{Slow}]{\text{OEt}^{\ominus}} \overset{\ominus}{C}H_2{-}CH_2{-}Br \xrightarrow{\text{Fast}} CH_2{=}CH_2$$

cation were formed it would react even in the absence of ethoxide ion, while in fact ethyl bromide is stable in the absence of base. The second is a more serious possibility, but it is excluded by the finding that no deuterium is incorporated in an alkyl bromide which is recovered from partial reaction in EtOD. If the proton is reversibly removed it must be replaced by deuterons from the solvent. The third mechanism is not excluded by this finding, since no equilibrium is suggested. It can be ruled out by the observation that base-catalyzed eliminations are run under conditions much too mild to permit formation of an unstabilized free carbanion at an appreciable rate. One may exclude any mechanism containing a step which would be slower than the observed over-all reaction rate. Simultaneous elimination appears to be the only reasonable mechanism.

Nevertheless, one might wonder whether everything really happens at precisely the same time. May not proton removal run slightly ahead of halide ion loss, so that a little negative charge builds up on carbon? Conversely, is it not possible that in some cases ionization of the leaving group may run ahead of proton loss, so that a little carbonium ion character is developed? The answer to these questions is "Yes." Cases are known in which E2 eliminations cover the whole range, from "almost carbanion" processes to "almost carbonium ion" processes. Evidence for this on the carbanion side comes most readily from a study of substituent effects, as in the *p*-substituted β-phenethyl derivatives. In Table 4–1 are listed the second-order rate constants for some substituted phenethyl bromides in an E2 elimination with sodium ethoxide in ethanol at 30°C. It can be seen that groups which could stabilize a carbanion by conjugation, such as the *p*-nitro group, strongly accelerate the reaction; groups which should destabilize a carbanion, such as the *p*-methyl group, slow the elimination compared with the unsubstituted case.

worse than

Substituent effects of this type have been put on a quantitative basis by Hammett. For each substituent, such as a *p*-methoxyl group, a substituent constant σ can be assigned. Thus $\sigma_{p\text{-}MeO}$ is – 0.268. Such a negative σ indicates that the *p*-methoxy group

TABLE 4–1 ▪

$$p\text{—R—}C_6H_4\text{—}CH_2\text{—}CH_2\text{—Br} \xrightarrow[\text{EtOH}]{\text{NaOEt}} p\text{—R—}C_6H_4\text{—}CH{=}CH_2$$

Rate = k_2 *(alkyl bromide)(OEt⁻)*

R	$k_2 \times 10^5$ (liters/mole second)
CH_3O—	16
CH_3—	23
H—	42
Cl—	191
CH_3CO—	1720
NO_2—	75,200

donates electrons to a benzene ring. The *p*-methyl group, a weaker electron donor, has a σ of only -0.170, while the electron-attracting *p*-cyano group has a positive σ, $+0.660$, and *p*-nitro is even more positive, $+0.778$. Each reaction considered in this treatment has a reaction constant ρ which measures the sensitivity of that reaction to substituent effects. Then the effect of the substituent on the rate constant is expressed with a simple equation.

$$\text{Log}\ (k_{\text{substituted}}/k_{\text{unsubstituted}}) = \sigma\,\rho$$

This, the Hammett equation, is a very general expression of the effect of aromatic ring substituents on side-chain reactivity. It can be applied to equilibria as well as rates. Thus the effect of ring substituents on the ionization constants of benzoic acids (the standard reaction, for which ρ is defined as 1.0) or of phenols (for which ρ is found to be 2.113, larger than 1.0 since the charge in a phenol anion is closer to the substituent than it is in a benzoate anion) can be correlated using the same set of σ constants for the various substituents. Since rates and acidities are covered by this equation, it bears some relation to the Brønsted catalysis law discussed in Chapter 2, and like the Brønsted law the Hammett equation describes a *linear free energy relationship*. Taft has defined a similar equation governing substituent effects in aliphatic systems, and in Chapter 5 we will see that the Hammett equation with slightly modified σ constants can be applied to aromatic substitution processes. However, for our present purposes it is enough to note that a constant ρ can be determined for a reaction by varying substituents and observing the effect of this on the rate. It is found that ρ for phenethyl elimination reactions depends on the particular leaving group.

For instance, with sodium ethoxide in ethanol at 30°C, *p*-cyanophenylethyl iodide is 23 times as reactive as is unsubstituted phenylethyl iodide.

$$\text{Log}\ 23 = 1.36 = (+\,0.660)\,\rho$$
$$\text{therefore}\ \rho = +\,2.07$$

The rates with other substituents also give this same value of ρ when I^- is the leaving group. In Table 4–2 are listed some relative rates and reaction constants for E2 eliminations of β-phenethyl compounds with various leaving groups; the reaction conditions are again sodium ethoxide in ethanol at 30°C.

The fact that ρ is positive for all these reactions means that they all develop some carbanion character, but at the transition state

TABLE 4–2 ▪

E2 Elimination of R—C_6H_4—CH_2—CH_2X

X	Relative rate (when R = H)	ρ
– I	26,600	+ 2.07
– Br	4,100	+ 2.14
– Toluenesulfonate	392	+ 2.27
– Cl	68	+ 2.61
– F	1	+ 3.12
– $\overset{+}{N}Me_3$	—	+ 3.77

this is more pronounced for the elimination of HF than for the elimination of HI. This is revealed by the fact that the HF elimination has a larger ρ, meaning that its transition state can be stabilized to a greater extent by groups which would stabilize a carbanion. Considering the reaction rates as well, these data agree with what one might expect for the different leaving groups. Thus, as the ethoxide ion begins to remove a proton from a phenethyl iodide some carbanion character develops, but iodide is such a good leaving group that it quickly begins to depart and the new double bond starts forming. Consequently only a moderate anionic charge ever develops on the carbon; this means that the substituent effects are limited in size. It also means that the reaction goes readily, since a full carbanion of this type would be of quite high energy and the transition state energy will be raised to the extent that anionic character must be developed. When fluoride ion is the leaving group more negative charge must develop before this ion can be ejected; this makes the reaction slower, and also more sensitive to substituent effects.

These data illustrate the fact that apparently very similar E2 eliminations can involve the development of varying amounts of carbanion character. This effect is sometimes invoked to explain the difference between "Saytzeff" and "Hofmann" orientation in elimination reactions, although steric hindrance effects are also involved to a major extent. E2 eliminations with neutral substrates, such as alkyl halides, alkyl toluenesulfonates, etc., ordinarily lead to the more substituted olefin. These cases are said to

$$CH_3-CH_2-\underset{CH_3}{\overset{Br}{C}}-CH_3 \xrightarrow[25^\circ]{NaOEt} \underset{72\%}{CH_3-CH=\underset{CH_3}{C}-CH_3} + \underset{28\%}{CH_3-CH_2-\underset{CH_3}{C}=CH_2}$$

follow the Saytzeff rule. On the other hand, E2 elimination in a quaternary ammonium hydroxide (Hofmann elimination) occurs with removal of the most acidic hydrogen. In simple cases this leads to formation of the least substituted olefin, since primary carbanions are more stable than secondary or tertiary carbanions (alkyl groups are electron donating relative to hydrogen atoms). This orientation is said to follow the Hofmann rule.

$$CH_3-CH_2-\overset{\oplus}{\underset{CH_2CH_2CH_3}{\overset{CH_2CH_3}{N}}}-CH_2CH_2CH_3 \xrightarrow[\Delta]{OH^-} \underset{96\%}{[CH_2=CH_2 + EtN(Pr)_2]} + \underset{4\%}{[CH_3CH=CH_2 + PrNEt_2]}$$

Hofmann elimination

The ρ values listed in Table 4–2 show that the transition state for Hofmann elimination has a considerable amount of carbanion character; given a choice the reaction which involves a better carbanion will occur. The transition state for HBr elimination has less carbanion character and more double-bond character, so in this case the reaction which forms the better olefin (alkyl groups stabilize olefins) will occur.

The stereochemistry of E2 eliminations supports the picture of more or less concerted elimination. Whenever possible the two groups eliminated assume a *trans*-coplanar position in the transi-

:B H C R′ R R′ R C X ≡ :B H R′ R R′ R X

Trans-coplanar Elimination

Cis-coplanar Elimination

tion state. The statement that "*trans*" elimination occurs refers to the positions of these groups, and not to the geometry of the resulting olefin. A few cases of *cis*-coplanar elimination have been found in compounds for which the *trans*-coplanar geometry is impossible.

E1cB Eliminations. In the cases so far discussed we have seen examples in which the leaving group ionized first (E1) or in which loss of the proton and the leaving group were more or less concerted (E2). The third possibility, ionization of the proton before loss of the leaving group, has been called the E1cB mechanism (elimination unimolecular conjugate base). It should be emphasized that kinetically such a process would be second order, since the formation of the carbanion would involve base as well as substrate, but the rate-determining step, loss of the leaving group, would be unimolecular.

$$-\overset{H}{\underset{|}{\overset{|}{C}}}-\overset{X}{\underset{|}{\overset{|}{C}}}- \underset{\text{Fast}}{\overset{B:}{\rightleftharpoons}} \overset{\ominus}{C}-\overset{X}{\underset{|}{\overset{|}{C}}}- \xrightarrow[\text{Slow}]{-X^{\ominus}} C=C$$

The E1cB mechanism

Such a process can only occur if the carbanion is strongly stabilized, and if the leaving group is sufficiently poor that it will not be lost from the developing anion by E2 elimination. Several cases are now known in which it can be demonstrated that deuterium exchange, via the carbanion, is faster than elimination.

This is observed in the base-catalyzed elimination of DF from (labeled) 1,1-dichloro-2,2-trifluoroethane.

$$CDCl_2-CF_3 \xrightarrow{OH^{\ominus}} {}^{\ominus}CCl_2-CF_3 \xrightarrow{\text{slow}} CCl_2=CF_2 + F^{\ominus}$$

$${}^{\ominus}CCl_2-CF_3 \xrightarrow[\text{fast}]{H_2O} CHCl_2-CF_3$$

However, the demonstration that a carbanion can be formed is not the same as proof that it is an intermediate in the elimination.

Thermal Eliminations. In contrast to these ionic eliminations, a number of processes are known in which elimination occurs by a unimolecular process without the attack of external reagents. Two synthetically useful examples are the formation of olefins by pyrolysis of esters or, at lower temperatures, xanthates.

$$\text{CH}_3\text{C(=O)O–C–C–H} \xrightarrow{500^\circ\text{C}} \text{>C=C<} + \text{CH}_3\text{CO}_2\text{H}$$

$$\text{CH}_3\text{SC(=S)O–C–C–H} \xrightarrow{300^\circ\text{C}} \text{>C=C<} + \left[\text{CH}_3\text{S–C(=O)–SH} \rightarrow \text{CH}_3\text{SH} + \text{O=C=S}\right]$$

Concerted mechanisms are written for these processes, supported by the fact that the reactions are carried out in the gas phase, where solvation of intermediate ions would be impossible, and by the fact that the reactions involve *cis* elimination. The measured entropy of activation, $\Delta S^{\ddagger}$, is negative for these processes, as expected for a cyclic transition state in which some of the initial freedom of rotation has been lost.

α-Eliminations

When chloroform is treated with strong base it loses HCl, forming dichlorocarbene. The carbene is not stable, but it has been trapped in various ways, e.g., by addition to olefins present during the carbene-forming reaction.

$$\text{HCCl}_3 + t\text{BuO}^{\ominus}\text{K}^{\oplus} \longrightarrow :\text{CCl}_2 + t\text{BuOH} + \text{KCl}$$

$$:\text{CCl}_2 \xrightarrow{\text{cyclohexene}} \text{7,7-dichlorobicyclo[4.1.0]heptane (C–Cl, Cl)}$$

Carbene formation can also be detected in other ways when chloroform is treated with aqueous base, and under these conditions fast hydrogen exchange occurs with the solvent, detectable if D_2O is the solvent. This base-catalyzed exchange shows that the trichloromethyl anion is formed reversibly; it is usually considered to be evidence that the elimination involves the anion. (However, as was mentioned above, demonstration that a carbanion can

$$HCCl_3 \underset{\text{Fast}}{\overset{OH^{\ominus}}{\rightleftarrows}} {}^{\ominus}CCl_3 \xrightarrow[\text{Slow}]{} :CCl_2 + Cl^{\ominus} \xrightarrow{\text{etc.}}$$

be formed is not really equivalent to demonstrating that it is an intermediate in the elimination.) The alternative concerted elimination mechanism has been found for the hydrolysis of chlorodifluoromethane.

$$OH^{\ominus} \; H{-}CF_2{-}Cl \longrightarrow H_2O + :CF_2 + Cl^{\ominus}$$

The change in mechanism occurs because of two factors: the trichloromethyl anion is more stable than the chlorodifluoromethyl anion, since chlorine can help stabilize the charge by use of its $3d$ orbitals, and difluorocarbene is more stable than dichlorocarbene

$$Cl_2\overset{\ominus}{C}{-}Cl \longleftrightarrow \overset{\ominus}{Cl}{=}C(Cl){-}Cl \longleftrightarrow \text{etc.}$$

since fluorine has $2p$ orbitals for π bonding with the vacant carbene $2p$ orbital, while chlorine must form the less stable $3p$—$2p$ π bond.

$$F{-}\ddot{C}{-}F \longleftrightarrow \overset{\oplus}{F}{=}\overset{\ominus}{C}{-}F \longleftrightarrow F{-}\overset{\ominus}{C}{=}\overset{\oplus}{F}$$

The addition of carbenes to olefins will be further discussed in Special Topic 4.

4–2 Addition to Carbon-Carbon Double Bonds

Almost all olefins will react with electrophilic reagents such as Br_2, while only compounds in which the double bond is activated by a carbonyl group or similar anion-stabilizing function will react with nucleophiles. Accordingly, electrophilic addition will be treated first.

Acid-catalyzed hydration. As discussed before, the mechanism of acid-catalyzed hydration of an olefin must be precisely the reverse of the acid-catalyzed dehydration of the alcohol under the same conditions. This reversible process may be pictured as follows.

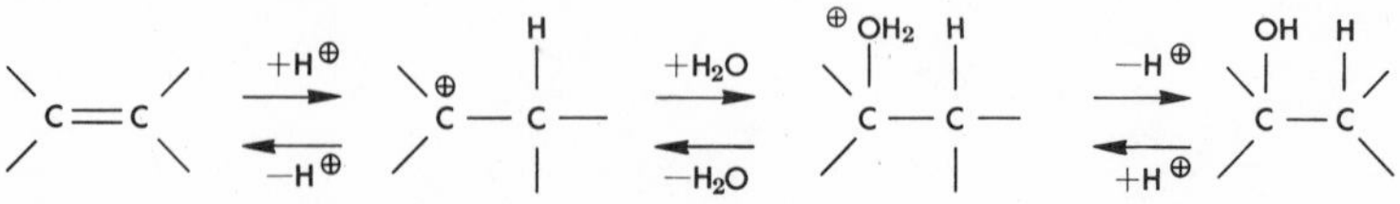

$$\mathrm{>C{=}C<} \underset{-H^{\oplus}}{\overset{+H^{\oplus}}{\rightleftharpoons}} \mathrm{>\overset{\oplus}{C}{-}\overset{H}{\overset{|}{C}}{-}} \underset{-H_2O}{\overset{+H_2O}{\rightleftharpoons}} \mathrm{>\overset{\oplus OH_2}{\overset{|}{C}}{-}\overset{H}{\overset{|}{C}}{-}} \underset{+H^{\oplus}}{\overset{-H^{\oplus}}{\rightleftharpoons}} \mathrm{>\overset{OH}{\overset{|}{C}}{-}\overset{H}{\overset{|}{C}}<}$$

The reverse will be recognized as an E1 elimination reaction, involving formation of the carbonium ion.

If protonation of the olefin were rapidly reversible, deuterium exchange should be observed between the olefin protons and deuterium in the solvent. However, 2-methyl-2-butene which is recovered after 50% hydration in D_2O contains no deuterium.

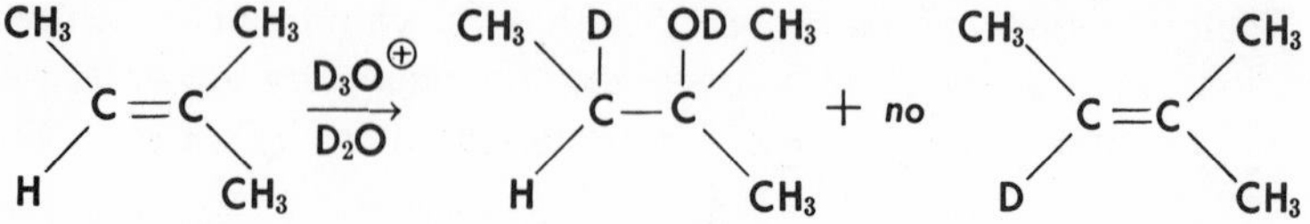

$$\mathrm{(CH_3)(H)C{=}C(CH_3)_2} \xrightarrow[D_2O]{D_3O^{\oplus}} \mathrm{(CH_3)(H)\overset{D}{\overset{|}{C}}{-}\overset{OD}{\overset{|}{C}}(CH_3)_2} + \textit{no}\ \mathrm{(CH_3)(D)C{=}C(CH_3)_2}$$

This shows that the carbonium ion is not reversibly formed. Thus the rate-determining step is carbonium ion formation, the latter rapidly going on to product alcohol. For other reasons there has been a suggestion that carbonium ion formation might involve two steps: reversible addition of a proton to the olefin to form a π complex, followed by slow collapse of this intermediate to the

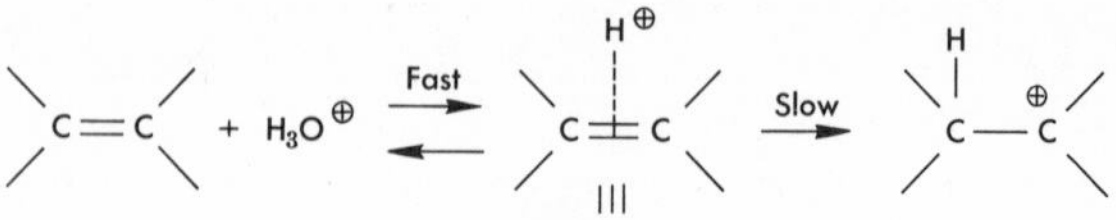

$$\mathrm{>C{=}C<} + \mathrm{H_3O^{\oplus}} \underset{}{\overset{\text{Fast}}{\rightleftharpoons}} \underset{\text{III}}{\mathrm{>C\overset{H^{\oplus}}{\overset{\vdots}{=}}C<}} \xrightarrow{\text{Slow}} \mathrm{>\overset{H}{\overset{|}{C}}{-}\overset{\oplus}{C}<}$$

$$\left[\ \mathrm{H}\ \ \mathrm{C{-}C}\ \right]^{\oplus}$$

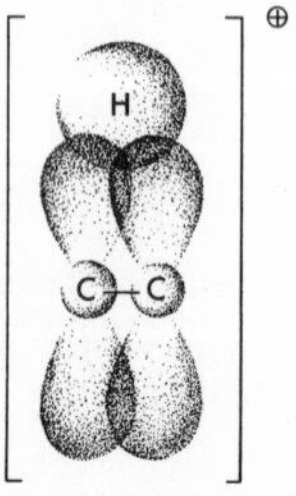

carbonium ion. The π complex could be pictured with overlap of the hydrogen 1*s* orbital and lobes of the carbon 2*p* orbitals. Evidence for π complexing in electrophilic aromatic substitution will be discussed in Special Topic 5.

Addition of HX. Of course, other nucleophiles may attack the carbonium ion as well, and the addition of HBr will have a mechanism similar to that outlined above. The stereochemistry of such additions is not yet clear. It is reported that addition of HBr to 1,2-dimethylcyclohexene occurs predominantly *trans,* while under other conditions the addition of DBr to acenaphthene has been found to involve *cis* stereochemistry.

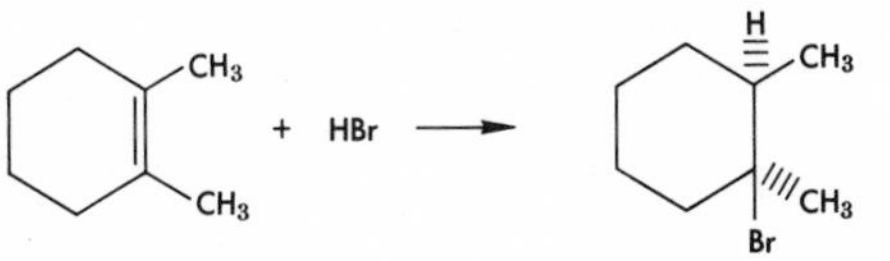

CH_3, CH_3 + HBr ⟶ H, CH_3, CH_3, Br — *Trans*-addition

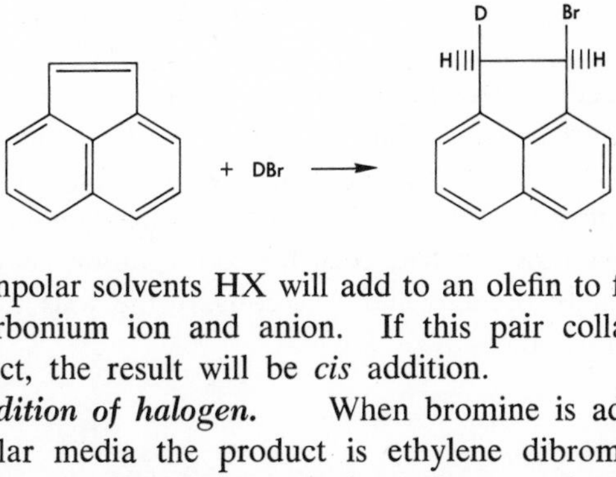

+ DBr ⟶ D, Br, H, H — *Cis*-addition

In nonpolar solvents HX will add to an olefin to form an ion pair of carbonium ion and anion. If this pair collapses rapidly to product, the result will be *cis* addition.

Addition of halogen. When bromine is added to ethylene in polar media the product is ethylene dibromide, but in the presence of chloride ion, of nitrate ion, or of other nucleophiles mixed adducts are obtained. Furthermore, in cases where the

$$CH_2{=}CH_2 + Br_2 \longrightarrow BrCH_2CH_2Br$$

$$CH_2{=}CH_2 + Br_2 \xrightarrow{Cl^-} BrCH_2CH_2Cl$$

$$CH_2{=}CH_2 + Br_2 \xrightarrow{NO_3^-} BrCH_2CH_2ONO_2$$

stereochemistry can be detected it is found that these additions occur *trans.* Since *trans* addition occurs even in noncyclic olefins

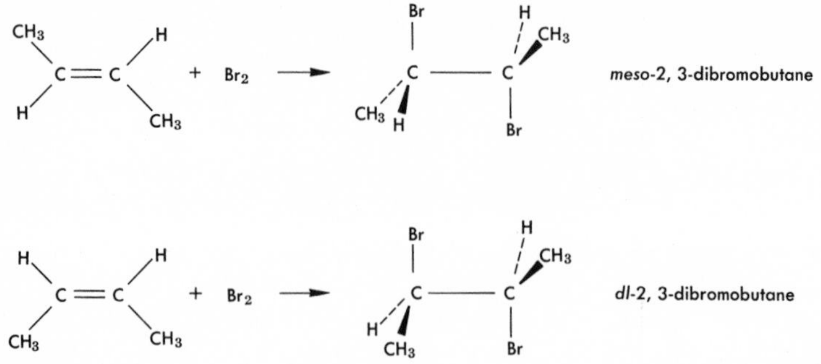

meso-2, 3-dibromobutane

dl-2, 3-dibromobutane

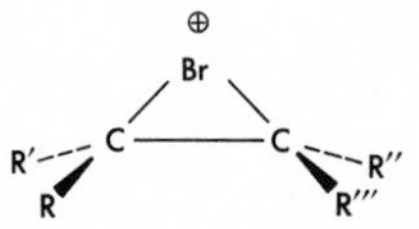

the formation of a free carbonium ion intermediate is excluded, and instead the participation of a bromonium ion is suggested. The bromonium ion can then be attacked by any nucleophile in the system, not merely by bromide ion, so the formation of mixed adducts is explained. More important, the stereochemistry is thus explained since displacement on the carbon of the bromonium ion should occur with inversion, leading to over-all *trans* addition, and since *cis*- and *trans*-olefins would form different bromonium ions. If a classical carbonium ion had been formed, rotation about the carbon-carbon single bond would interconvert the ions derived from *cis*- and *trans*-butene.

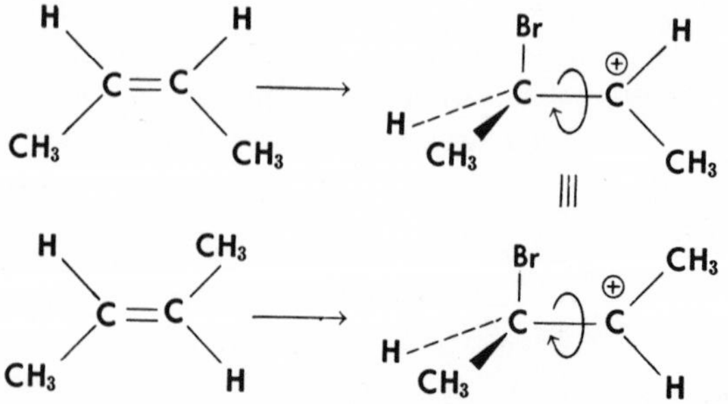

Attack on the bromonium ion by a nucleophile could occur at either carbon. Usually the nucleophile attacks the more substituted carbon, showing that the substitution has much of the character of an S_N1 reaction favoring the better carbonium ion.

However, with *t*-butylethylene, attack at the secondary carbon is so hindered that substitution occurs at the primary carbon.

$$(CH_3)_3C{-}CH{=}CH_2 \xrightarrow[CH_3OH]{Br_2} (CH_3)_3C{-}CHBr{-}CH_2OCH_3$$

It must not be concluded that bromonium (chloronium, iodonium) ions are always formed in halogen additions. If the classical carbonium ion can be strongly stabilized it may be the preferred intermediate. This is revealed in the *cis*-addition of chlorine to acenaphthene, in contrast to the usual *trans* additions. In this case a classical carbonium ion is probably formed paired with a chloride ion, and collapse of the ion pair occurs with over-all *cis* addition.

Hydroboration. H. C. Brown has developed a variety of synthetic procedures which start with the addition of diborane, B_2H_6, to olefins. The reagent acts as if it were a source of BH_3, a Lewis acid with six-electron boron which undergoes electrophilic attack on a double bond. Internal hydride transfer occurs in the intermediate (I) so the overall result is *cis* addition of BH_3 to the olefin. With an excess of olefin the alkylborane formed (e.g., II) may react as did BH_3 to form *di*- or *tri*-alkylboranes.

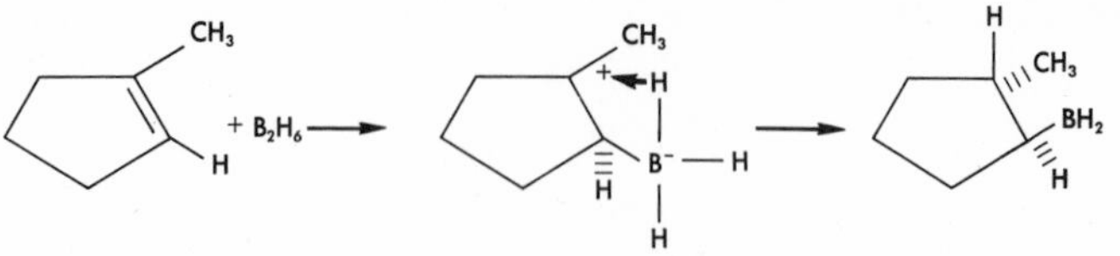

$$\text{1-methylcyclopentene} + B_2H_6 \longrightarrow \text{I} \longrightarrow \text{II}$$

The alkylboranes may be converted to various other products. One of the most useful procedures is oxidation with alkaline hydrogen peroxide, producing boric acid and an alcohol.

$$BR_3 + 3H_2O_2 \xrightarrow{NaOH} 3ROH + B(OH)_3$$

The borane coordinates with H_2O_2, and the intermediate undergoes a reaction related to the perbenzoate rearrangement discussed in Special Topic 3.

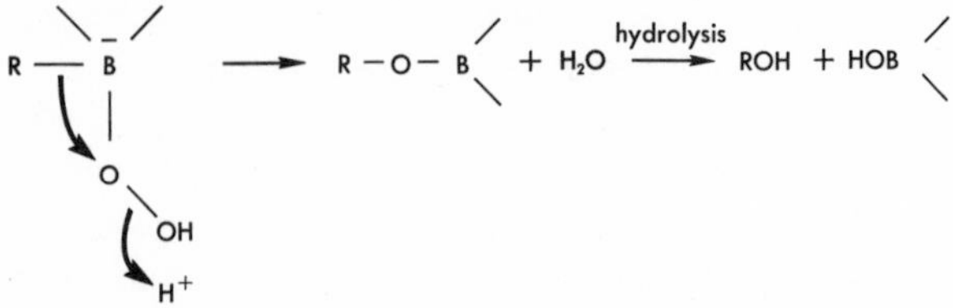

$$R-O-B\!< + H_2O \xrightarrow{\text{hydrolysis}} ROH + HOB\!<$$

The migrating group in such rearrangements retains its configuration, and the subsequent hydrolysis of the borate ester involves cleavage of the B—O bond. Thus in the product alcohol the OH group has the same configuration as the original boron atom in the borane; e.g., oxidation of II leads to III; it should be noted that the over-all addition of H_2O to a double bond by this sequence is not only stereospecific but also leads to an orientation the reverse of that from acid-catalyzed hydration.

$$\text{II} \xrightarrow[NaOH]{H_2O_2} \text{III}$$

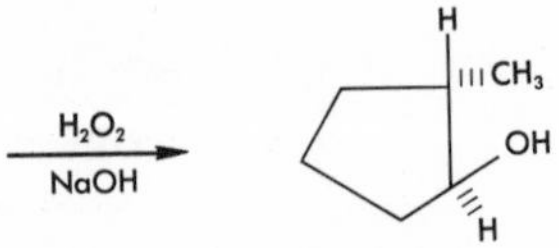

Nucleophilic addition to olefins. Just as electrophilic addition to olefins may be considered the reverse of E1 elimination, nucleophilic additions to olefins are the reverse of eliminations involving carbanions. The nucleophilic addition of HY to a double bond may be symbolized as follows.

$$\rangle C{=}C\langle + Y^{\ominus} \rightarrow -\underset{Y}{\overset{|}{C}}-\overset{\ominus}{C}\langle \xrightarrow{H^+} -\underset{Y}{\overset{|}{C}}-\underset{H}{\overset{|}{C}}-$$

In general such a process will require two special features: the nucleophile must be a good one, and the resulting carbanion must be strongly stabilized. It will be seen that these are similar to the factors cited for the operation of the E1cB elimination mechanism: a stabilized carbanion, and a poor leaving group (which will probably therefore be a good nucleophile). The two mechanisms are connected by the principle of microscopic reversibility; if an elimination reaction involves an E1cB mechanism, then its reverse under the same conditions must involve nucleophilic addition via the carbanion.

As was discussed earlier, some elimination reactions which appear to have "carbanion character" actually involve extremes of the E2 elimination mechanism, in which proton removal runs ahead of departure of the leaving group. Similarly, the microscopic reverse of such an elimination would involve attack by the nucleophile but then the beginning of proton donation before carbanion development was complete.

$$\rangle C{=}C\langle \;(Y^{\ominus},\ H{-}OH) \longrightarrow \rangle\underset{Y}{C}-\overset{\delta^{-}\cdots H\cdots OH^{\delta-}}{C}\langle \longrightarrow \rangle\underset{Y}{C}-\overset{H}{C}\langle \;\; {}^{\ominus}OH$$

Thus while a number of nucleophilic additions to activated double bonds are usually written as if the free carbanion were involved, the possibility must be kept in mind that proton addition could occur before complete carbanion formation.

A few examples of such reactions will be encountered in Chapter 6.

$$C_6H_5CH{=}C(CN)(CO_2^{\ominus}) \xrightarrow[H_2O]{KCN} C_6H_5CH(CN){-}\overset{\ominus}{C}(CN)(CO_2^{\ominus}) \longrightarrow C_6H_5CH(CN){-}CH(CN)(CO_2^{-})$$

or

$$C_6H_5CH(CN){-}\overset{\delta\ominus}{C}(CN)(CO_2^{\ominus}) \cdots H \cdots \overset{\delta\ominus}{OH}$$

$$\text{Rate is found} = k\,[\text{Olefin}]\,[CN^{-}]$$

General References

D. Banthorpe, *Elimination Reactions* (Elsevier Publishing Co., New York, 1963). A monograph in which a wide range of mechanisms are critically examined.

D. J. Cram, "Olefin-Forming Elimination Reactions," in M. Newman, ed., *Steric Effects in Organic Chemistry* (John Wiley & Sons, New York, 1956). A good survey of mechanisms with some emphasis on stereochemistry.

J. F. Bunnett, "The Mechanism of Bimolecular β-Elimination Reactions," *Angewandte Chemie, International Edition,* **1,** 225 (1962). A review article, in English, with many references.

J. Hine, *Physical Organic Chemistry* (McGraw-Hill Book Co., New York, 1962). Chapters 8 and 9 discuss both eliminations and additions. Chapter 22 covers α-eliminations.

W. Saunders, "Elimination Reactions in Solution," in S. Patai, ed., *The Chemistry of Alkenes* (John Wiley & Sons, New York, 1964). A critical treatment at a rather advanced level.

4

Special Topic

▪ THREE- AND FOUR-CENTER ADDITION REACTIONS

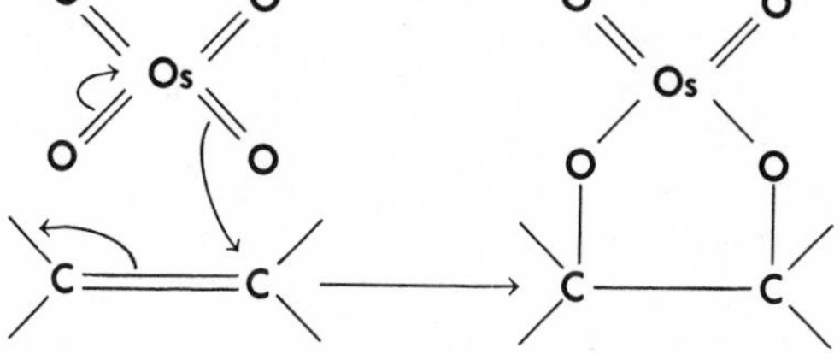

THERE ARE A NUMBER of reactions in which a reagent adds to a double bond by simultaneous attack on both unsaturated carbons, or where this is at least a formal possibility. For instance, in the reaction of an olefin with osmium tetroxide to form a cyclic osmate ester there is no need to postulate any sort of charged intermediate. The over-all change can be pictured as involving a simple cyclic flow of electrons, with a simultaneous valence change of the osmium atom. The Diels-Alder reaction[1] can also be written with a cyclic flow of electrons, without the necessity to postulate transient intermediate ions.

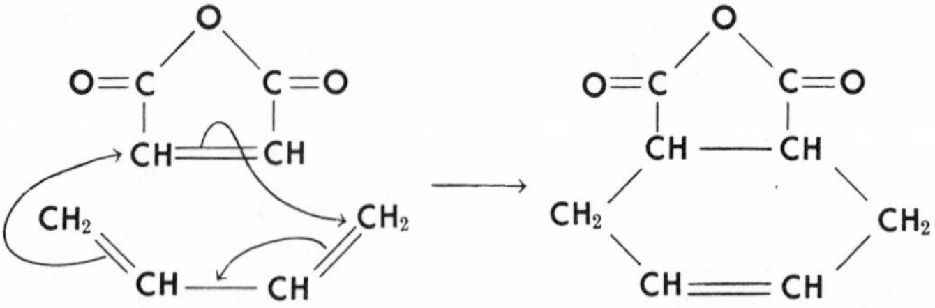

The addition of a carbene to an olefin is an example of a three-center process, in which again no unstable intermediate is formally required.

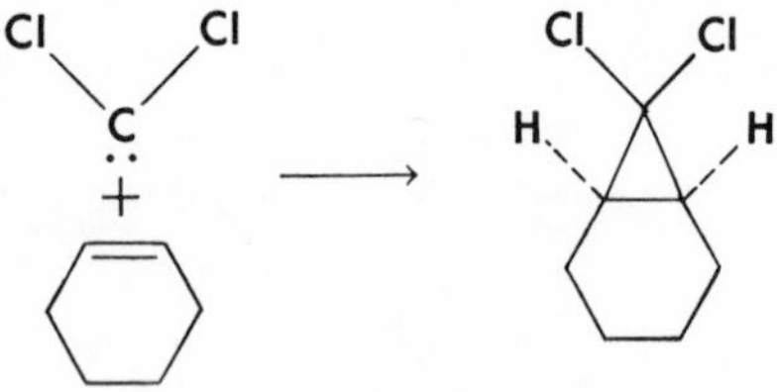

The question we will examine is the following: in such reactions, is addition to the double bond really simultaneous, or are the two bonds to carbon formed one at a time?

1,1-Addition Reactions

The addition of a carbene to an olefin is considered a 1,1-addition, since both new bonds come to the same carbon (of the carbene). Similarly, formation of a bromonium ion is a 1,1-addition, as is epoxidation by peracids, although in these cases another fragment is simultaneously eliminated.

1. (a) M. Kloetzel, "The Diels-Alder Reaction with Maleic Anhydride," in R. Adams, ed., *Organic Reactions,* Vol. 4 (John Wiley & Sons, New York, 1948), p. 1; (b) H. Holmes, "The Diels-Alder Reaction: Ethylenic and Acetylenic Dienophiles," in R. Adams, ed., *Organic Reactions,* Vol. 4 (John Wiley & Sons, New York, 1948), p. 60; (c) L. Butz and A. Rytina, "The Diels-Alder Reaction: Quinones and Other Cyclenones," in R. Adams, ed., *Organic Reactions,* Vol. 5 (John Wiley & Sons, New York, 1949), p. 136; (d) S. Needleman and M. Chang Kuo, "Diels-Alder Synthesis with Heteroatomic Compounds," *Chemical Reviews,* **62,** 405 (1962).

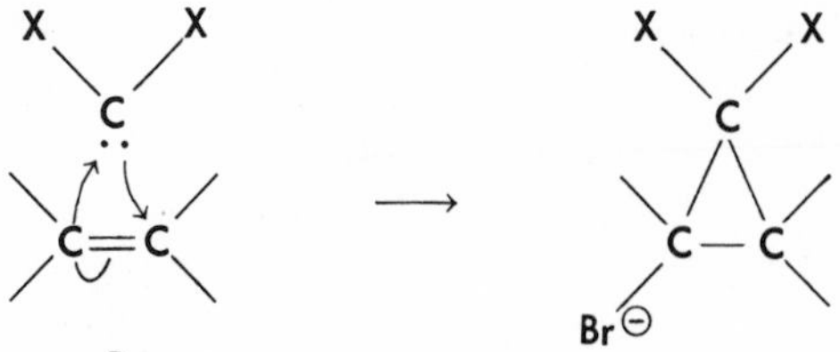

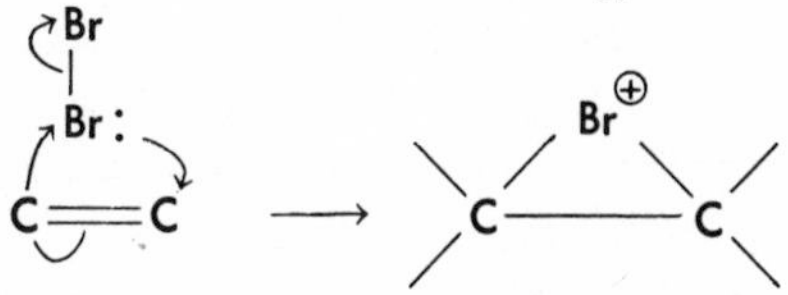

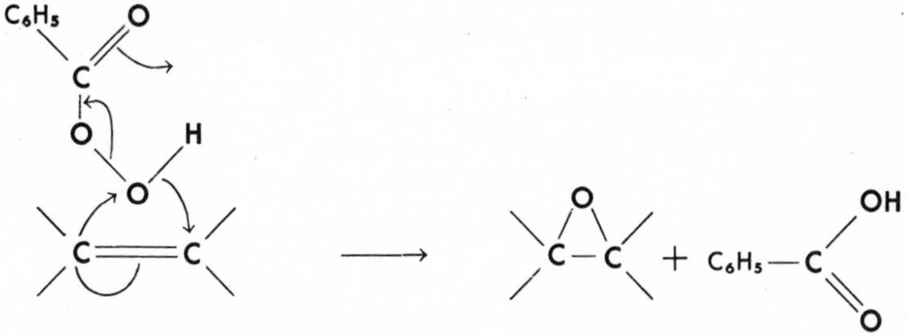

These processes have been written as if the addition were indeed simultaneous; the evidence in favor of this is the finding that the addition is stereospecific. As has already been discussed for bromonium ion formation, reactions with *cis*-2-butene and with *trans*-2-butene would lead to the same mixture of products if addition were stepwise. It is actually found that the isomeric olefins yield isomeric products in bromonium ion reactions, in epoxidation,[2] and in some carbene additions.[3]

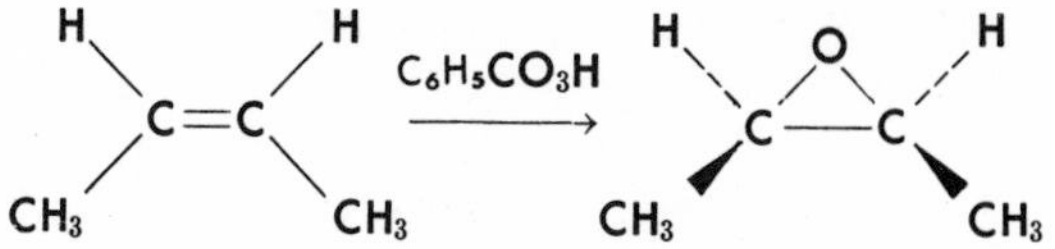

2. E. Gould, *Mechanism and Structure in Organic Chemistry* (Henry Holt & Co., New York, 1959), p. 534.
3. (a) P. Skell and A. Garner, "The Stereochemistry of Carbene-Olefin Reactions," *Journal of the American Chemical Society,* **78,** 3409 (1956); (b) R. Woodworth and P. Skell, "Methylene, CH_2. Stereospecific Reaction with *cis*- and *trans*-2-Butene," *Journal of the American Chemical Society,* **81,** 3383 (1959).

Not all carbenes undergo stereospecific addition to olefins. There are two possible electronic states[4] for any carbene, the singlet state and the triplet state; apparently only the singlet gives stereospecific addition. Considering CH_2 itself, in the singlet state (in which all electrons are paired) the carbon is approximately sp^2 hybridized. Two of the sp^2 orbitals are used for single bonding to hydrogen while the third contains the unshared pair of electrons (actually spectroscopic evidence[5] suggests that this orbital has more *s* character than the other two). The remaining *p* orbital is vacant. Thus the singlet state resembles a carbonium ion, except that a proton is missing. In triplet $:CH_2$ the carbon is (approximately) *sp*-hydridized, each of the *sp* hybrid orbitals being used for the C—H single bonds;[6] the two unshared electrons on carbon are placed in the p_y and p_z orbitals left over. By Hund's rules[7] one of these electrons will go into each *p* orbital, so that they stay as far apart as possible, and they will also unpair spins so as to give a "triplet state," i.e., a state with two unpaired spins.

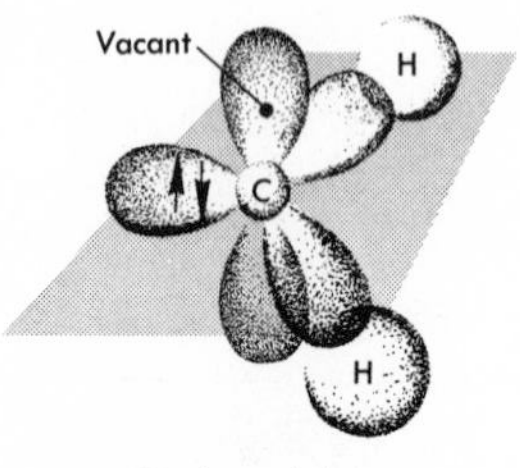

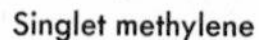
Singlet methylene

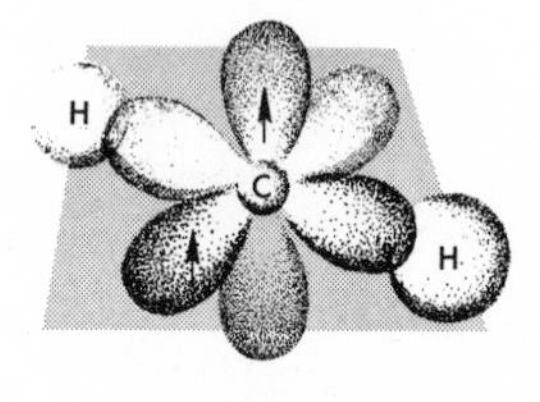

Triplet methylene

In the triplet methylene the two unshared electrons are in *p* orbitals, while in the singlet they are in an sp^2 hybrid orbital, which

4. For a good review, cf. (a) J. Hine, *Physical Organic Chemistry* (2nd ed., McGraw-Hill Book Co., New York, 1962), Chapter 24; (b) J. Hine, *Divalent Carbon* (Ronald Press, New York, 1964); (c) W. Kirmse, *Carbene Chemistry* (Academic Press, New York, 1964).
5. Ref. 4 (a), p. 493.
6. Cf. ref. 5, but also E. Wasserman, A. Trozzolo, and W. Yager, "ESR Hyperfine of Randomly Oriented Triplets: Structure of Substituted Methylenes," *Journal of Chemical Physics,* **40,** 2408 (1964), who conclude that triplet carbenes are not completely linear.
7. C. Coulson, *Valence* (2nd ed., Oxford University Press, London, 1961), p. 36.

has more *s* character and thus lower energy. Thus one might have expected the singlet state to be more stable. However, the decrease in electron repulsion from separating the two electrons in different *p* orbitals and from unpairing their spins apparently more than makes up for the higher energy of a *p* orbital, so that the triplet state of $:CH_2$ is more stable than the singlet state.

When diazomethane is decomposed by heat or light, singlet-state methylene is produced (unless a special type of photosensitizer is used, cf. Chapter 8). If the decomposition is carried out in the presence of an olefin, stereospecific addition of this singlet methylene occurs.[8] Even though the triplet state of $:CH_2$ is more stable, it is relatively difficult to interconvert singlet and triplet

$$\overset{\ominus}{C}H_2-\overset{\oplus}{N}\equiv N \xrightarrow[\text{Heat}]{\text{Light or}} (\uparrow\downarrow)C(H)H \xrightarrow{CH_3(H)C=C(H)CH_3} \text{cis-1,2-dimethylcyclopropane}$$

states; addition occurs before there is time for the initially formed singlet to be converted to the more stable triplet. However, in the gas phase in high dilution there is enough time for decay from the singlet to the triplet states[9]; furthermore, using certain types of photosensitizers[10] it is possible to produce triplet-state methylene directly from diazomethane. Then it is found that addition of this triplet-state methylene is no longer stereospecific.

$$CH_2N_2 \xrightarrow{\text{Light, benzophenone as photosensitizer}} H-\overset{\uparrow}{C}\!\uparrow\!-H \xrightarrow{\text{Cis-2-butene}} \text{cis-1,2-dimethylcyclopropane} + \text{trans-1,2-dimethylcyclopropane}$$

$$CH_2N_2 \xrightarrow{\text{Light}} (\uparrow\downarrow)C(H)H \xrightarrow{\text{Gas phase in } N_2} H-\overset{\uparrow}{C}\!\uparrow\!-H$$

8. Ref. 3b.
9. F. Anet, R. Bader, and A. van der Auwera, "Chemical Evidence for a Triplet Ground State for Methylene," *Journal of the American Chemical Society,* **82,** 3217 (1960).
10. K. Kopecky, G. Hammond, and P. Leermakers, "The Triplet State of Methylene in Solution," *Journal of the American Chemical Society,* **84,** 1015 (1962).

The difference in behavior between these two electronic states of a carbene is not surprising. The singlet adds in one step, but addition of the triplet to a double bond is a two-step process; the first step leads to a triplet diradical, which must then be converted (slowly) to the singlet state before ring closure can occur. As expected from this mechanism, olefins which are particularly

$$H_2\overset{\uparrow}{C}\uparrow + C\overset{\uparrow\downarrow}{-}C \longrightarrow -\overset{\uparrow CH_2}{\overset{|}{C}}-\overset{\uparrow}{C}- \longrightarrow -\overset{CH_2\downarrow}{\overset{|}{C}}-\overset{\uparrow}{C}- \longrightarrow \text{cyclopropane } (CH_2)$$

reactive toward free radical additions are very reactive toward the triplet states of carbenes,[11] although they do not exhibit any such preference for the singlet carbenes.

When methylene iodide is treated with zinc (as a zinc-copper alloy) in the presence of an olefin, CH_2 is added to the double bond.[12] At one time this was considered to be a carbene addition, but it is now clear that the intermediate is a "carbenoid," i.e., a carbene-like fragment bonded to something else, in this case zinc iodide. It is also clear that a number of other "carbenes" produced in solution by elimination reactions, even ":CCl_2" from the alkaline

$$CH_2I_2 + Zn \longrightarrow ZnI-CH_2-I \xrightarrow{\;>C=C<\;} >C\overset{CH_2}{\frown}C< + ZnI_2$$

hydrolysis of chloroform, are not completely free.[13] Such carbenoids exist in the singlet state and give one-step stereospecific addition.

11. R. Etter, H. Skovronek, and P. Skell, "Diphenylmethylene, a Diradical Species," *Journal of the American Chemical Society,* **81,** 1008 (1959).
12. F. Blanchard and H. Simmons, "Cyclopropane Synthesis from Methylene Iodide, Zinc-Copper Couple, and Olefins. II," *Journal of the American Chemical Society,* **86,** 1337 (1964).
13. Cf. W. Miller and D. Whalen, "Trichloromethyllithium, an Electrophilic Reagent," *Journal of the American Chemical Society,* **86,** 2089 (1964), and references therein.

1,2-Addition Reactions[14]

On being heated to 200°C, tetrafluoroethylene dimerizes to octafluorocyclobutane. This is considered a 1,2-cycloaddition reaction, in this case a symmetrical one. Fluorinated olefins also will add 1,2 to other materials, e.g., butadienes, and similar cycloadditions are known for ketenes, allenes, and activated olefins such as acrylonitrile.

(1) $CF_2{=}CF_2 \xrightarrow{200^\circ C}$ CF_2-CF_2 / CF_2-CF_2 (octafluorocyclobutane ring)

(2) $CH_2{=}C(CH_3)-C(CH_3){=}CH_2 + CCl_2{=}CF_2 \xrightarrow{100^\circ C} CH_2{=}C(CH_3)-C(CH_3)$ ring with CH_2, CF_2, CCl_2

(3) cyclopentadiene $+ CH_2{=}C{=}O \xrightarrow{100^\circ C}$ bicyclic cyclobutanone (O)

(4) $CH_2{=}C{=}CH_2 \xrightarrow{400^\circ C}$ 1,2-dimethylenecyclobutane (85%) + 1,3-dimethylenecyclobutane (15%)

(5) $CH_2{=}CHCN \xrightarrow{250^\circ C}$ 1,2-dicyanocyclobutane (CN, CN) + 1,2-dicyanocyclobutane stereoisomer (CN, CN)

In all these except the first reaction there is in principle a choice of orientation, and the actual direction of addition is that shown. Addition cannot be a simple, completely concerted process, since such a mechanism would not explain the preferred orientations. Considering reaction (5), for instance, the dimerization of acrylonitrile, it seems that both electrostatic and steric effects would favor the formation of 1,3-dicyanocyclobutane. The product

14. J. Roberts and C. Sharts, "Cyclobutane Derivatives from Thermal Cycloaddition Reactions," in A. C. Cope, ed., *Organic Reactions,* Vol. 12 (John Wiley & Sons, New York, 1962), p. 1.

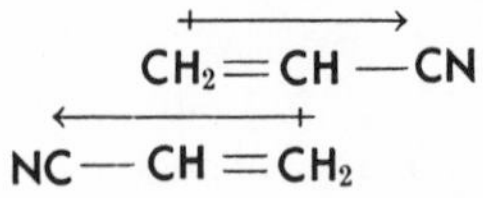

$$CH_2{=}CH{-}CN$$
$$NC{-}CH{=}CH_2$$

actually formed is 1,2-dicyanocyclobutane. For such reasons it has long been considered that in these cycloadditions one bond is formed more rapidly than the other. In this case a diradical intermediate may be postulated, since it is known that radicals are stabilized by conjugation with cyano groups.

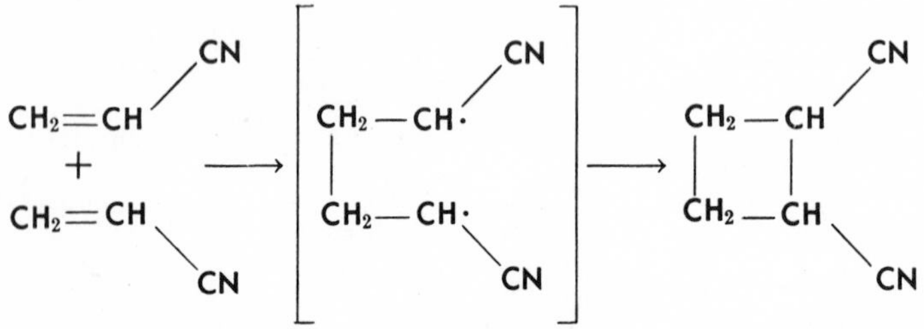

$$\begin{matrix} CH_2{=}CH{-}CN \\ + \\ CH_2{=}CH{-}CN \end{matrix} \longrightarrow \left[\begin{matrix} CH_2{-}\dot{C}H{-}CN \\ | \\ CH_2{-}\dot{C}H{-}CN \end{matrix} \right] \longrightarrow \begin{matrix} CH_2{-}CH{-}CN \\ |\quad\;\; | \\ CH_2{-}CH{-}CN \end{matrix}$$

An ionic intermediate would not explain this orientation since a cyano group does not stabilize adjacent positive charge, although either a diradical or a zwitterionic intermediate structure could account for the ketene orientation in reaction (3). One of the

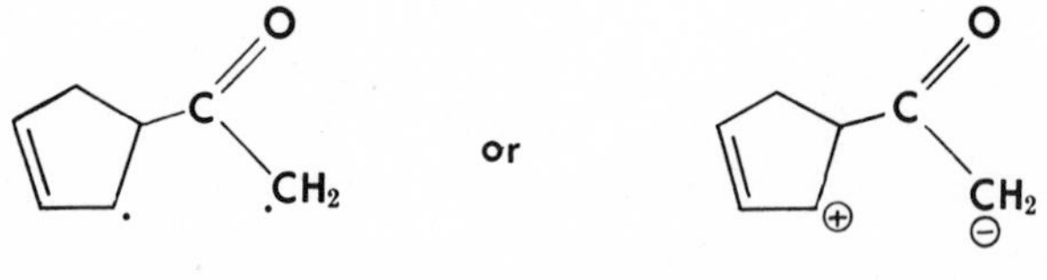

[Structures: cyclopentenyl radical–C(=O)–$\dot{C}H_2$ or cyclopentenyl cation (⊕)–C(=O)–$CH_2^{\ominus}$]

questions one might ask about such intermediates is the following: Do they represent really free diradicals, perhaps with unpaired electron spins, or do they simply involve a very long bond in which the paired electrons are so far apart that they behave to some extent as if they were in free radicals? The answer to this question is not known for all cases, but recent experiments[15] related

15. P. Bartlett *et al.*, "Cycloaddition," *Journal of the American Chemical Society,* **86,** 616, 622, 628 (1964), *Journal of Organic Chemistry,* **32,** 1290 (1967); cf. also S. Proskow, H. Simmons, and T. Cairns, "Stereochemistry of the Cycloaddition Reaction . . . ," *Journal of the American Chemical Society,* **85,** 2341 (1963), for a different result.

to reaction (2) show that for this case a real diradical is involved, not simply a long bond.

Reaction of 1,1-dichloro-2,2-difluoroethylene with *cis,cis*-2,4-hexadiene yields two stereoisomeric products, I and II. These are formed because the intermediate diradical has time to rotate about the new single bond before it collapses to product; if formation of one bond simply ran slightly ahead of the other, such isomerization would not be possible. Thus even though in the starting material the two hydrogen atoms were *cis* to each other, in product I they are *trans* because of this opportunity for single-bond rotation in the intermediate.

$CH_3CH{=}CHCH{=}CHCH_3 + CF_2{=}CCl_2 \rightarrow$ diradical (CF_2—$\dot{C}Cl_2$) $\xrightarrow{\text{single-bond rotation}}$ I + II

I: cyclobutane (CF_2—CCl_2) with CH_3, H, CH=CHCH$_3$ substituents

II: cyclobutane (CF_2—CCl_2) with CH_3, CH=CHCH$_3$, H substituents

1,3-Addition Reactions

A number of reagents give 1,3-addition to olefins. The reaction of olefins with OsO_4 has already been mentioned. Ozone also undergoes 1,3-addition to double bonds, although the "initial" ozonide then rearranges to a more stable ozonide.[16]

The concept of 1,3-additions as a related class of reactions was

16. P. Bailey, "The Reaction of Ozone with Organic Compounds," *Chemical Reviews,* **58,** 925 (1958). For evidence that some ozonolyses have more complex mechanisms see R. Murray, R. Youssefyeh, and P. Story, "Ozonolyis . . . ," *Journal of the American Chemical Society,* **89,** 2429 (1967) and references therein.

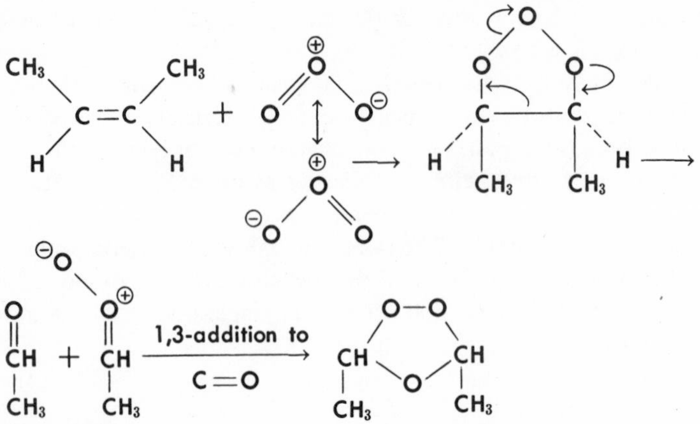

1,3-addition to C=O

first advanced in 1938.[17] However, activity in discovering new reactions of this type and in investigating mechanisms has been quite recent, and is chiefly due to the efforts of Huisgen and his collaborators.[18] 1,3-Dipolar systems may be symbolized generally as III or IV; diazomethane is an example of system III, while ozone is an example of system IV.

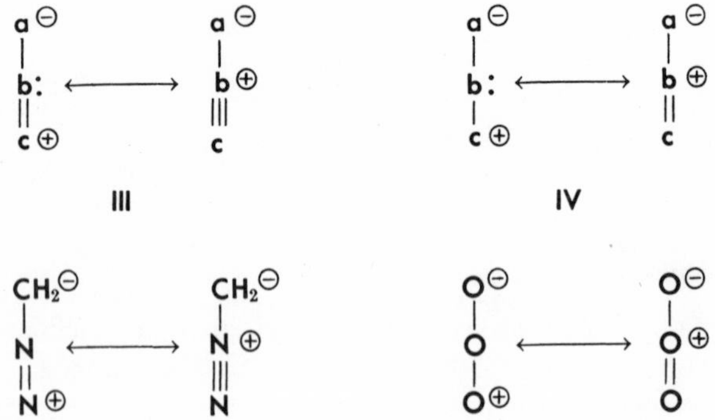

III

IV

17. L. Smith, "Systems Capable of Undergoing 1,3-Additions," *Chemical Reviews,* **23,** 193 (1938).
18. (a) R. Huisgen, "1,3-Dipolar Cycloadditions," *Angewandte Chemie, International Edition,* **2,** 563 (1963); (b) R. Huisgen, "Kinetics and Mechanism of 1,3-Dipolar Cycloadditions," *Angewandte Chemie, International Edition,* **2,** 633 (1963).

Most of the possible systems III and IV are known, where a, b, and c are carbon, nitrogen, or oxygen. Many of them will undergo 1,3-additions to double bonds, and the evidence so far[18] suggests that these additions are more or less concerted (i.e., the second bond at least begins to form before the first is complete).

The primary evidence for this is stereospecificity in all additions so far examined. Thus diazomethane affords different products with the esters of dimethylfumaric and of dimethylmaleic acids, and unstable diphenylnitrilamine (V) affords stereospecific adducts with the stilbenes if it is generated in their presence.

$\overset{\ominus}{C}H_2-N=\overset{\oplus}{N}$ + $(CH_3)(CH_3O_2C)C=C(CH_3)(CO_2CH_3)$ ⟶ pyrazoline: CH_2, N, N; CH_3, C, CO_2CH_3; C, CH_3, CO_2CH_3

$\overset{\ominus}{C}H_2-N=\overset{\oplus}{N}$ + $(CH_3)(CH_3O_2C)C=C(CO_2CH_3)(CH_3)$ ⟶ pyrazoline: CH_2, N, N; CH_3, C, CO_2CH_3; C, CO_2CH_3, CH_3

$C_6H_5C(Cl)=N-NHC_6H_5$ $\xrightarrow{Et_3N}$ $[C_6H_5-\overset{\oplus}{C}=N-\overset{\ominus}{N}-C_6H_5]$ V

V + $(C_6H_5)(H)C=C(C_6H_5)(H)$ ⟶ $C_6H_5-C=N-N-C_6H_5$; C_6H_5, C, H; C, C_6H_5, H

V + $(C_6H_5)(H)C=C(H)(C_6H_5)$ ⟶ $C_6H_5-C=N-N-C_6H_5$; C_6H_5, C, H; C, H, C_6H_5

Such stereospecificity shows that a free open-chain intermediate is not formed.

Nonetheless, 1,3-dipolar addition is probably not a completely concerted process. The best evidence for this comes from the effect of substituents on the rates of some addition reactions.[18] Simple olefins are much less reactive than are olefins substituted with conjugating groups; for additions by benzonitrile oxide in ether at 20°C, for instance, the following relative rates are found.

$CH_2{=}CH{-}$alkyl	$CH_2{=}CH{-}$phenyl	$CH_2{=}CH{-}CO_2R$
3.9	14	100

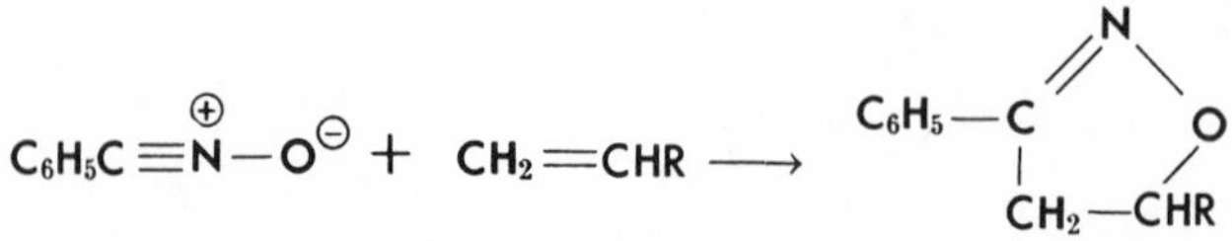

$$C_6H_5C{\equiv}\overset{\oplus}{N}{-}O^{\ominus} + CH_2{=}CHR \longrightarrow \text{(ring: } C_6H_5{-}C{=}N{-}O{-}CHR{-}CH_2{-}\text{)}$$

If addition had been completely concerted one might have expected no particular help from conjugating groups, while with an intermediate radical or anion being partially developed in the reacting olefin the acceleration can be accounted for.

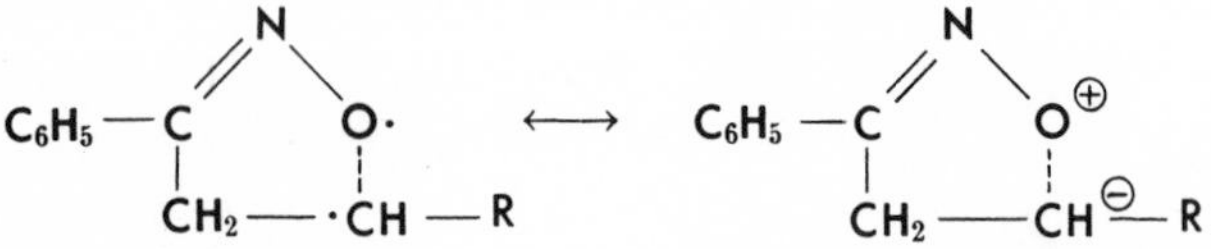

$$C_6H_5{-}C(=N{-}O\cdot)\ CH_2{-}\cdot CH{-}R \longleftrightarrow C_6H_5{-}C(=N{-}O^{\oplus})\ CH_2{-}CH^{\ominus}{-}R$$

Addition to the olefin results in a particular orientation, as was shown above. This might also be taken as an indication that the transition state has some charge-separated character, but other evidence[18] suggests that steric effects are mostly involved, addition to the olefin occurring so as to separate the bulky substituent groups as much as possible. In general, 1,3-dipolar additions are not accelerated by polar solvents; therefore in the transition state there is no more charge separation than in the starting materials.

1,4-Addition Reactions

In the Diels-Alder reaction[1] a conjugated diene reacts across its 1,4-positions with an olefin; the latter is called the dienophile in this reaction. Almost any olefin may serve as a dienophile under some conditions, but olefins substituted by electron-withdrawing conjugating groups are usually more reactive. The reaction of methyl *trans*-crotonate with cyclopentadiene is a typical Diels-Alder reaction.[19]

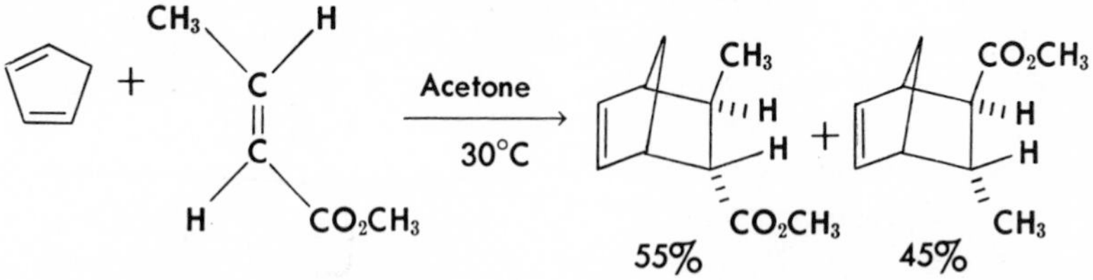

It will be noted that in the two products stereospecific addition has occurred across the crotonic ester bond, i.e., the methyl and carbomethoxyl groups are still *trans*. Thus addition to the double bond has been *cis;* many cases have been examined, and this result is always found. Accordingly, we can again say that the addition must be more or less concerted, and no free diradical intermediate capable of rotation about the new single bond is involved.

In contrast to this evidence that the two bonds are simultaneously formed, the *orientation* in unsymmetrical Diels-Alder additions can be most easily understood in terms of open-chain intermediates.[20] Thus the reaction of 2-methoxybutadiene with acrolein yields the adduct expected if an intermediate zwitterion or diradical had been formed, since the methoxyl and carbonyl groups would

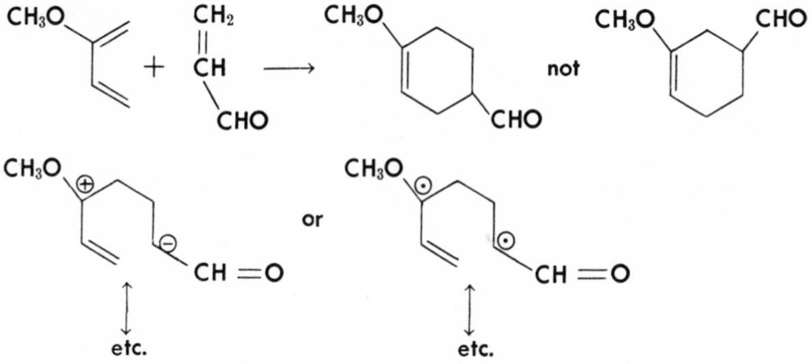

19. J. G. Martin and R. K. Hill, "Stereochemistry of the Diels-Alder Reaction," *Chemical Reviews,* **62,** 537 (1962).
20. Cf. R. Woodward and T. Katz, "The Mechanism of the Diels-Alder Reaction," *Tetrahedron,* **5,** 70 (1959).

stabilize such intermediates. Although this is still a matter of controversy,[21] it seems most likely that no such intermediate is involved, but that the two single bonds are forming at the same time, with one running ahead of the other in most cases. The transition state for the above reaction can be represented with one short bond and one long bond. In a long bond the two electrons may both be near one carbon or they may be more or less localized with one on each carbon (though still with spins paired). In some cases the transition state will thus appear to be dipolar, while for other reactions substituent effects will suggest diradical character. With this concept of a "long bond" both the stereospecificity and the substituent effects can be explained.

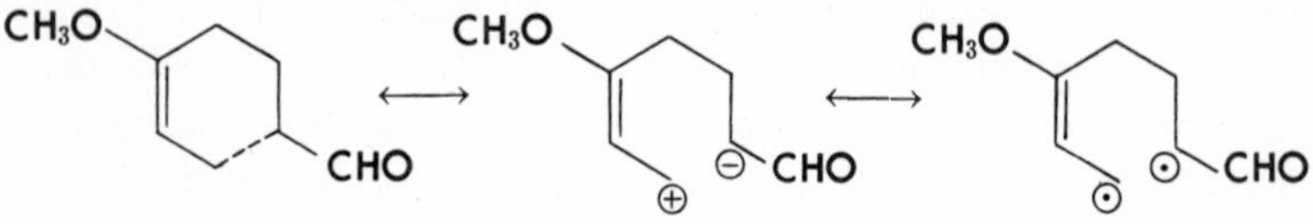

It is striking that the Diels-Alder reaction seems to be concerted, in the sense that both new bonds are forming at essentially the same time, while at least some of the 1,2 additions we discussed earlier occur one bond at a time. Since nonconcerted reactions require the formation of unstable intermediates, we might wonder why any cycloaddition would be nonconcerted. Is there any reason why two ethylene molecules cannot simply come together in a single-step reaction to form cyclobutane? Of course, such a process has a considerable entropy disadvantage since four atoms must be aligned, but this entropy problem is also found in the Diels-Alder process which is in fact concerted.

The distinction in mechanisms seems to derive chiefly from the fact that the Diels-Alder reaction involves the cyclic flow of six electrons, while dimerization of ethylene involves four electrons. In the transition state for a concerted process these electrons become delocalized over the system before they are localized again in the new bonds, and transition states with $4n + 2$ cyclically delocalized electrons are much better than those with $4n$ cyclically delocalized electrons. This sounds like the distinction between aromatic and antiaromatic systems discussed in Special Topic 1,

21. J. Berson and A. Remanick, "The Mechanism of the Diels-Alder Reaction," *Journal of the American Chemical Society*, **83**, 4947 (1961). Provides a good critical discussion.

and it has the same origins in molecular quantum mechanics. A consideration of *orbital symmetries* also shows what the problem is in olefin dimerization (or other 2 + 2 cycloadditions, involving two electrons from each component).

It is necessary to consider a molecular orbital description of the two new single bonds in cyclobutane formed by dimerization of ethylene.[22]

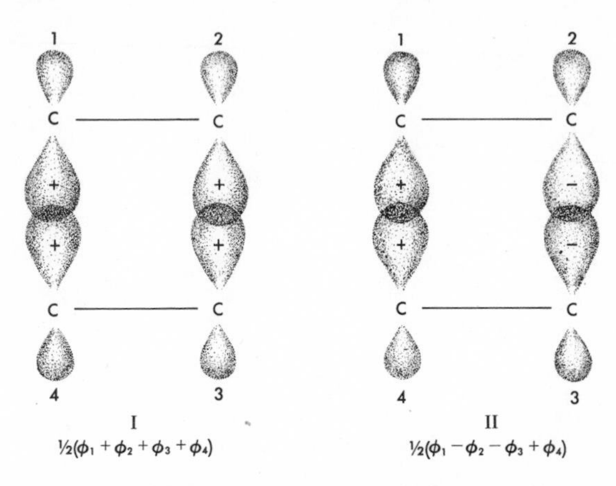

I $\frac{1}{2}(\phi_1 + \phi_2 + \phi_3 + \phi_4)$

II $\frac{1}{2}(\phi_1 - \phi_2 - \phi_3 + \phi_4)$

The σ bonds are formed by overlap of lobes of the sp^3 hybrid orbitals; as with p orbitals, the function corresponding to one lobe may have either a positive or negative sign, and bonding occurs when lobes of two overlapping sp^3 orbitals have the same sign. The four sp^3 orbitals on carbons 1, 2, 3, and 4 may be combined in two ways, I and II, so as to have bonding between 1 and 4, and 2 and 3. The (+) or (−) signs have no intrinsic significance; a structure like I with all (−) signs would be equivalent to I, signifying simply that all four sp^3 lobes have the *same* sign. Four electrons are involved in the two new bonds; two electrons go into m.o. I and two into m.o. II. Each of these is a *molecular* orbital, involving all four atoms. Thus orbital I has not only the desired 1–4 and 2–3 bonding interactions, but also some extra 1–2 and 3–4 π-like bonding. However, in II the 1–2 and 3–4 interactions are *antibonding,* and cancel the extra π-like bonds in I. The net result with two electrons in I and two electrons in II is equivalent to the normal cyclobutane description with localized bonds.

22. R. Hoffmann and R. B. Woodward, "Selection Rules for Concerted Cycloaddition Reactions," *Journal of the American Chemical Society,* **87,** 2046 (1965).

Since for most purposes the two-electron localized bond description of σ bonds is equivalent to a full m.o. description, we have not used the latter up until now. In the present case, however, we are concerned with the transition state in a hypothetical concerted cycloaddition of two ethylenes. The two ethylenes can themselves be lined up in two ways, considering orbital symmetries.

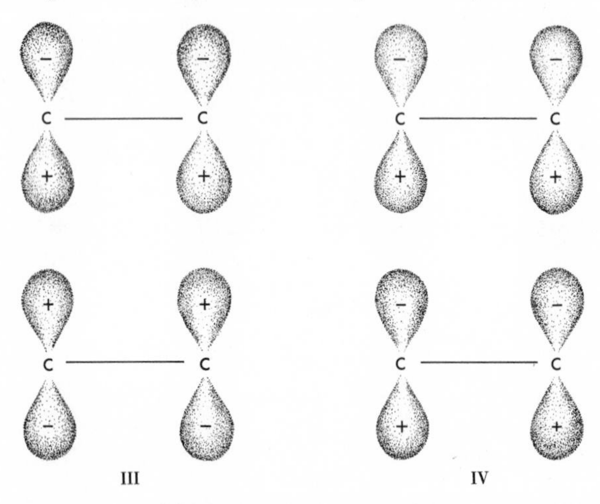

Arrangement III would go smoothly over to I of the cyclobutane as the reaction proceeded, but arrangement IV would not go to II but instead to a very high energy structure in which the two new σ interactions are *antibonding*. In fact, the arrangement which could go to II is V, in which the two ethylenes themselves have not bonds but antibonds. Thus one of the two cyclobutane ground-state m.o.'s is derived from a strongly excited state of the two ethylenes, and *vice versa*. *For the transition state of a concerted addition to be of low energy the electrons at the transition state must occupy orbitals intermediate between ground-state orbitals of the starting material and ground-state orbitals of the product.* In this case no such intermediate orbitals are possible, but in the 2 + 4 Diels-Alder reaction they are.

This latter point is most easily seen if we consider molecular symmetry. Ethylene has a plane of symmetry perpendicular to the carbon-carbon bond axis, and relative to this symmetry plane the two p_z orbitals may be either symmetric (same sign and magnitude, leading to the π bond) or antisymmetric (same magnitude but opposite sign, leading to the π antibond). These are symbolized **S** and **A.**

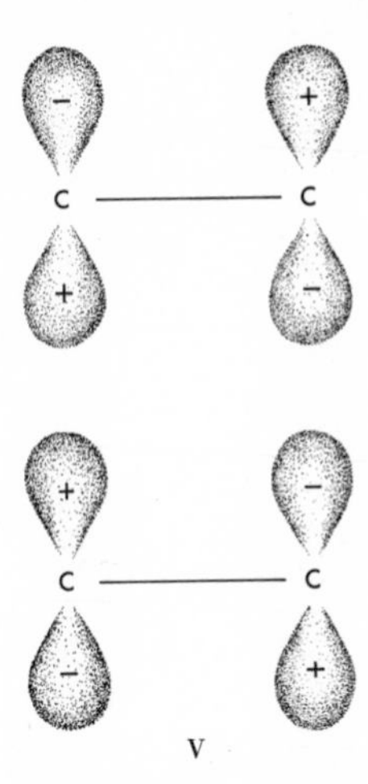

V

When two ethylenes concertedly close to a cyclobutane, this same molecular symmetry plane is always maintained, and any wave function must therefore be either **S** or **A** with respect to the plane. In cyclobutane, for instance, orbital I is **S** and orbital II is **A** with respect to the plane (which is perpendicular to both the 1–2 and 3–4 single bonds). However, in the two starting ethylene arrangements, III and IV are both **S.** There is no way in which an **S** orbital in the starting material and an **A** orbital in the product could be smoothly interconverted through some intermediate function. Either the intermediate wave function changes its sign or it doesn't on crossing a symmetry plane, so it must be either **S** or **A.**

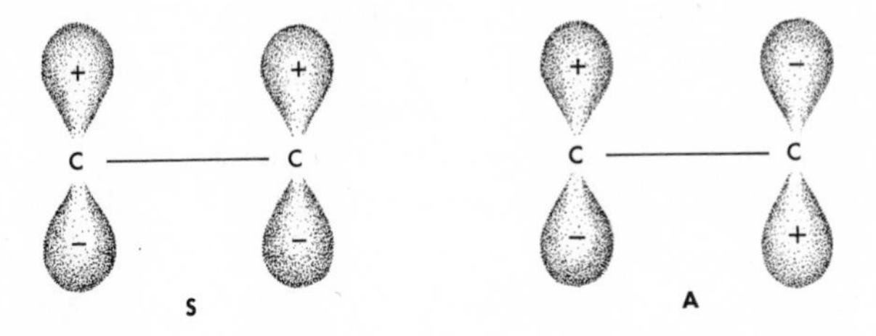

In the Diels-Alder reaction between an olefin and a diene the situation is different. Butadiene has two occupied π m.o.'s, which are **S** and **A** with respect to a symmetry plane. (See next page.) The product cyclohexene has three new bonds to accommodate the six electrons involved in the change: a new π bond and two new σ bonds. The π bond is **S** relative to the symmetry plane (as is

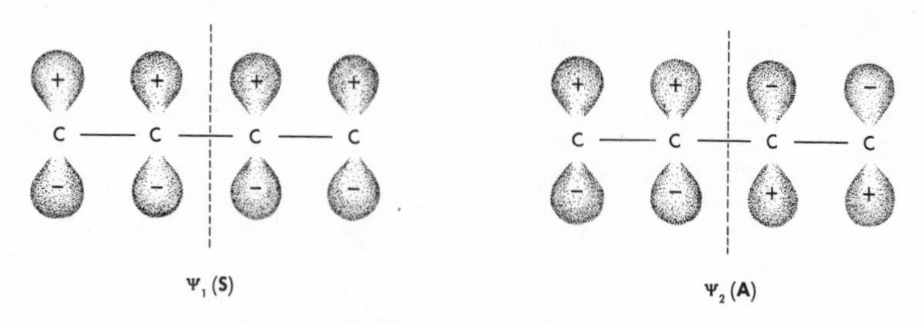

the reacting bond of the ethylene molecule), while the two σ bonds can be combined into an **S** and an **A** molecular orbital.

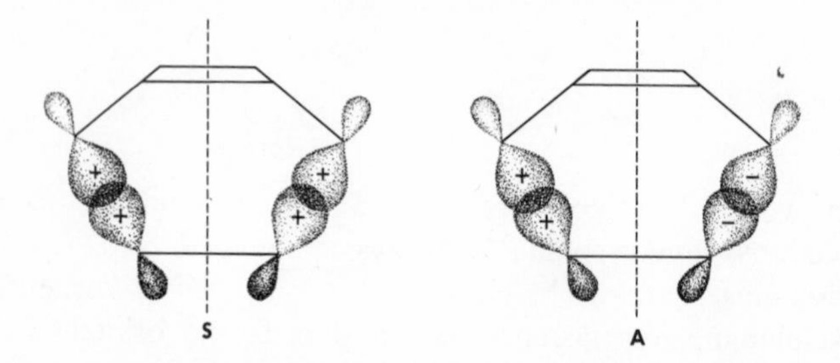

The six electrons of the starting materials (olefin and diene) are thus in two **S** and one **A** orbitals with respect to the symmetry plane, and so are the six electrons in the product. Orbitals retain their symmetries throughout the reaction so the transition state can resemble both starting materials and product. It can have some of the bonding characteristics of both, and thus be stabilized.

This type of argument will be pursued in more detail in Special Topic 8.

5

AROMATIC SUBSTITUTION

AS WAS DISCUSSED in Special Topic 1, there are many types of aromatic systems. Nonetheless, the chemistry of benzene and its simple derivatives has been studied in the most detail, so we shall concern ourselves with reactions on a benzene ring. Nitration of benzene and conversion of chlorobenzene to aniline illustrate the two major classes of aromatic substitution reactions; the first is an electrophilic substitution by nitronium ion, while the second is a nucleophilic substitution by the amide anion. As we will see, in

$$C_6H_6 + NO_2^{\oplus} \longrightarrow C_6H_5NO_2 + H^{\oplus}$$

$$C_6H_5Cl + NH_2^{\ominus} \longrightarrow C_6H_5NH_2 + Cl^{\ominus}$$

neither case does the actual mechanism involve a one-step displacement, however. Free radical substitution reactions are also known, but they will not be considered here.

5–1 Electrophilic Substitution

Three questions are usually of concern in aromatic substitution reactions: (1) What is the attacking species; (2) how does it carry out the substitution; and (3) how is the reaction influenced by other groups on the benzene ring? We shall consider the first two questions for several important substitution reactions, and shall deal with the third question at the end of this section.

Nitration. Benzene may be nitrated by a mixture of nitric and sulfuric acids. Physical evidence clearly shows that the nitronium ion, NO_2^+, is formed under these conditions. The earliest type of evidence was the finding from freezing-point studies that one molecule of nitric acid forms four particles in sulfuric acid.

$$HNO_3 + 2\,H_2SO_4 \longrightarrow NO_2^{\oplus} + H_3O^{\oplus} + 2\,HSO_4^{\ominus}$$

More recently, spectroscopic studies have demonstrated the presence of NO_2^+ in such solutions. A variety of evidence indicates that NO_2^+ is the actual nitrating agent with the ordinary H_2SO_4—HNO_3 mixture. Thus, it is possible to prepare authentic salts of nitronium ion, such as $NO_2^+\ BF_4$, and such salts will nitrate aromatic compounds. Furthermore the relative reactivities of different aromatic compounds toward authentic nitronium salts are the same as their relative reactivities toward nitric-sulfuric acid mixture, suggesting that the same species is attacking in both cases.

A number of mechanisms could be imagined for attack by nitronium ion. Direct displacement of hydrogen is perhaps the most obvious, but it has been ruled out by examination of the *isotope effect.* Studies of benzene and its derivatives containing deuterium or tritium show that such hydrogen isotopes are replaced as easily as is protium—ordinary hydrogen. However, whenever a proton is lost in the rate-determining step of a reaction it is found that there is an isotope effect, the heavier deuterium or tritium being lost less readily. The absence of an isotope effect here shows that the proton is not lost in the rate-determining step of nitration, and helps establish the two-step mechanism. The removal of proton by some base in the system (e.g., HSO_4) must be a fast step; if it were rate-determining we would again expect a large isotope effect.

In Special Topic 5, the mode of attack of NO_2^+ will be discussed in more detail. There it will also be seen that under some conditions nitrations do not involve the free nitronium ion.

Sulfonation. Treatment of benzene with fuming sulfuric acid affords benzenesulfonic acid. In this case SO_3 (or HSO_3^+) attacks the aromatic ring in the first step, but the attack is apparently reversible. This is revealed by the fact that there is a large isotope effect in this case, so proton removal is rate-determining. From this evidence alone, proton loss might be simultaneous with attack by SO_3, but the two-step mechanism seems more likely.

Halogenation. Many aromatic compounds can be brominated with solutions of bromine in acetic acid, and kinetic studies support a mechanism involving attack by molecular bromine.

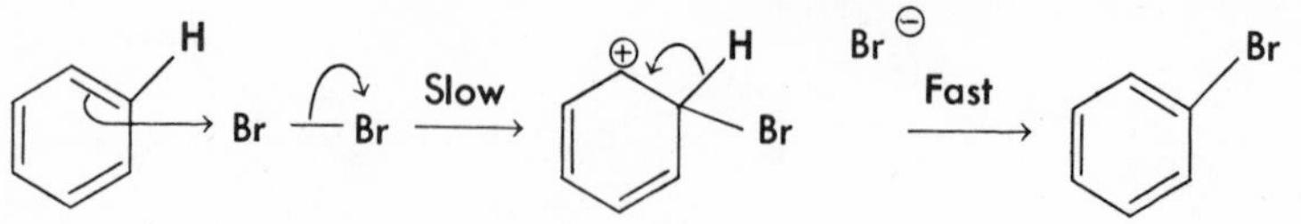

More detailed studies of aromatic halogenation have been performed in aqueous solution (although preparative reactions are usually not performed in this solvent).

When chlorine is dissolved in water a number of equilibria are set up involving such species as hypochlorous acid (HOCl), hypochlorous acidium ion (H_2OCl^+), and hypochlorite ion (ClO^-). The relative concentrations of these species depends on the pH and the chloride ion concentration.

$$\underline{Cl_2} + H_2O \rightleftharpoons \underline{H_2OCl^{\oplus}} + Cl^{\ominus}$$

$$H_2OCl^{\oplus} \rightleftharpoons \underline{HOCl} + H^{\oplus}$$

$$HOCl \rightleftharpoons ClO^{\ominus} + H^{\oplus}$$

The three species underlined have all been identified as responsible for aromatic chlorination in some cases, by a study of the effect of pH and chloride ion concentration on the rate of chlorination of some substrate. Of these, Cl_2 is apparently the most reactive, and it will attack unactivated benzene rings. Hypochlorous acid is quite unreactive, since the aromatic compound must displace a hydroxide ion from the chlorine atom, but it has been identified as the species which chlorinates phenoxide ion, a very reactive aromatic species.

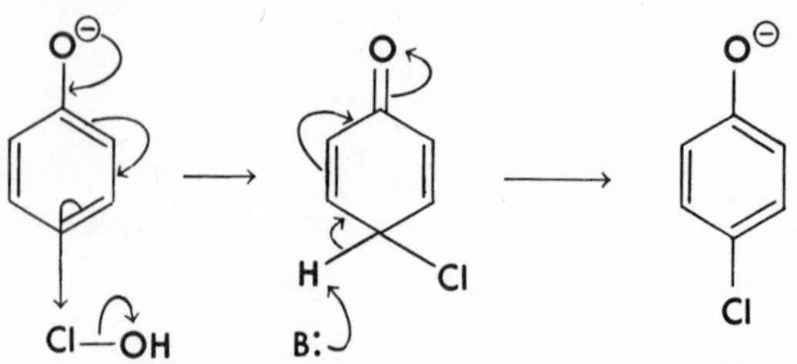

Under unusual conditions Cl^+ can also apparently be formed, and it is the most reactive chlorinating species of all.

Friedel-Crafts reactions. Treatment of an aromatic compound with an alkylating agent leads to substitution, the so-called Friedel-Crafts alkylation. The most common alkylating agents are mixtures of alkyl halides, such as methyl bromide, and strong Lewis acids such as aluminum bromide. The reaction may be considered a nucleophilic substitution on the methyl bromide by the aromatic ring, for which there are two obvious possible mechanisms. Either the aluminum bromide removes the bromide ion first, leaving a methyl cation to attack the ring, or there is an S_N2 displacement on the methyl group and the bromide ion is removed simultaneously.

$$CH_3Br + AlBr_3 \overset{?}{\rightleftarrows} CH_3^{\oplus} + AlBr_4^{\ominus} \longrightarrow \text{(ring with } CH_3 \text{, } H \text{, } \oplus\text{)} \longrightarrow \text{(ring with } CH_3\text{)}$$

or

$$\text{(benzene)} \; CH_3 - Br \text{---} AlBr_3 \longrightarrow \text{(ring with } \oplus \; H \text{, } CH_3\text{)} \longrightarrow \text{(ring with } CH_3\text{)}$$

Since the methyl cation is so unstable, we would be surprised if the first mechanism were correct. The direct displacement mechanism has been proved for this reaction by the finding that methyl iodide gives a different mixture of substitution products on toluene than does methyl bromide. If the methyl cation were the attacking species it should behave in the same way no matter how it is formed.

$$\text{(toluene, } CH_3\text{)} + CH_3X + AlBr_3 \longrightarrow \text{(ring with } CH_3\text{, } CH_3 \text{ ortho)} + \text{(ring with } CH_3\text{, } CH_3 \text{ meta)} + \text{(ring with } CH_3\text{, } CH_3 \text{ para)}$$

CH_3I	:	49%	11%	40%
CH_3Br	:	54%	17%	29%

With alkyl groups which afford more stable cations, such as *t*-butyl, it is the carbonium ion itself that attacks the aromatic ring. Carbonium ion rearrangements in the alkyl group are frequently observed.

Direct evidence for the proposed reaction intermediates in aromatic substitution has been obtained by Olah using NMR spectroscopy. A mixture of mesitylene with an alkyl halide and Lewis acid at low temperature yielded the substitution intermediate I. This went on to the final product at a higher temperature.

$$\text{mesitylene} + CH_3CH_2F + BF_3 \xrightarrow{-80°C} \text{I} \xrightarrow{-15°C} \text{ethylmesitylene}$$

$$\text{I} = \text{1-H-1-}CH_2CH_3\text{-2,4,6-tri-}CH_3\text{-benzenium ion} \quad BF_4^{\ominus}$$

Friedel-Crafts acylation of aromatic rings involves reaction with a Lewis acid and an acid chloride or other acylating agent. In this case there is evidence that the halide ion is lost first, the attacking species being an acylium ion.

$$R{-}\overset{O}{\overset{\|}{C}}{-}Cl + AlCl_3 \rightleftharpoons R{-}\overset{\oplus}{C}{=}O \longleftrightarrow R{-}C{\equiv}\overset{\oplus}{O} \qquad AlCl_4^{\ominus}$$

$$C_6H_6 + R{-}\overset{\oplus}{C}{=}O \longrightarrow [C_6H_6{-}C(=O)R]^{\oplus} \longrightarrow C_6H_5{-}C(=O)R$$

Substituent effects. Since the transition state in electrophilic aromatic substitution involves the creation of positive charge on the aromatic ring, substitution is facilitated by groups which can help stabilize such a charge. These groups affect the rate at which different aromatic compounds can be attacked, and they also affect the position of attack in a given compound such as toluene. In simple qualitative terms a group such as the methyl in toluene is said to be *activating,* since toluene is more reactive than benzene, and *ortho-para directing,* since preferential substitution occurs in these positions, as the above Friedel-Crafts alkylation showed.

These two different kinds of reactivity have been collected in the concept of a "partial rate factor." Instead of asking "How reactive is toluene compared with benzene, and what is the percentage of ortho, meta, and para substitution" for some reaction, we ask "How reactive is the para position of toluene compared with a position on benzene?" This relative reactivity would be the *partial rate factor.* Thus for bromination in acetic acid the partial rate factor for the position in toluene para to the methyl—p_f^{Me}—is 2420. The partial rate factor for meta substitution—m_f^{Me}—is 5.5.

Simple resonance arguments make it clear why the para position in toluene is over 2,000 times as reactive as a position in benzene. The transition state resembles the charged intermediate, and in one of the resonance forms there is a stable tertiary carbonium ion instead of a secondary one.

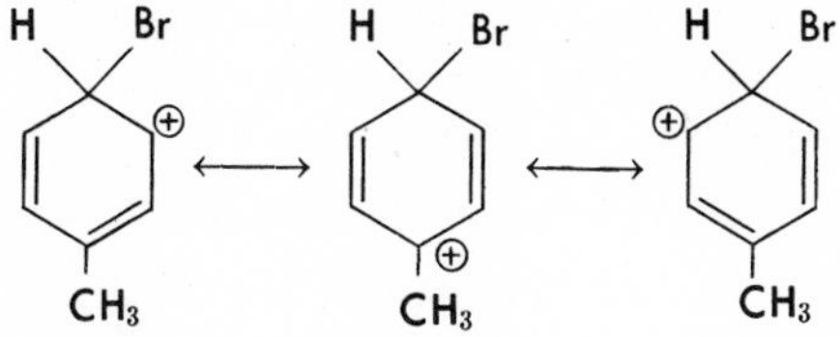

H Br ⊕ ⟷ H Br ⊕ CH3 ⟷ H Br ⊕

CH_3 CH_3 CH_3

For meta substitution all three resonance forms are secondary carbonium ions, so meta substitution is much less favorable.

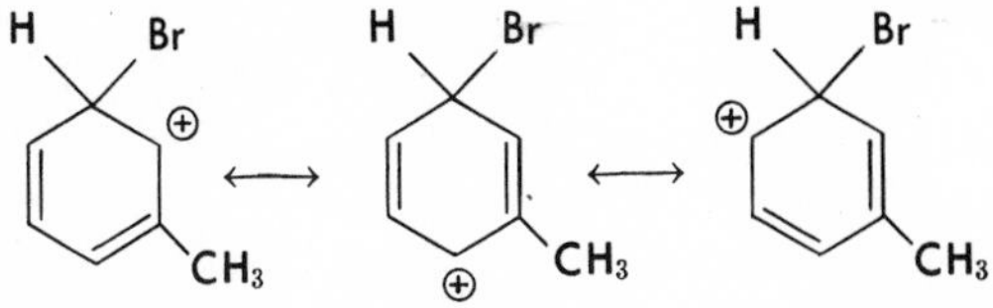

H Br ⊕ CH_3 ⟷ H Br ⊕ CH_3 ⟷ H Br ⊕ CH_3

Each meta position is only 5.5/2420 times as reactive as the para position, although a meta position in toluene is still 5.5 times as reactive as a benzene position. This latter increase must reflect a small inductive stabilization of the transition state by methyl even when it is not directly located on a charged carbon.

The partial rate factor for bromination of toluene in an ortho position is 600. Here we would again expect strong activation because one of the resonance forms is a tertiary carbonium ion. The fact that an ortho position is only ¼ as reactive as the para position is due to steric hindrance (although there is 50% as much ortho as para product since there are two ortho positions).

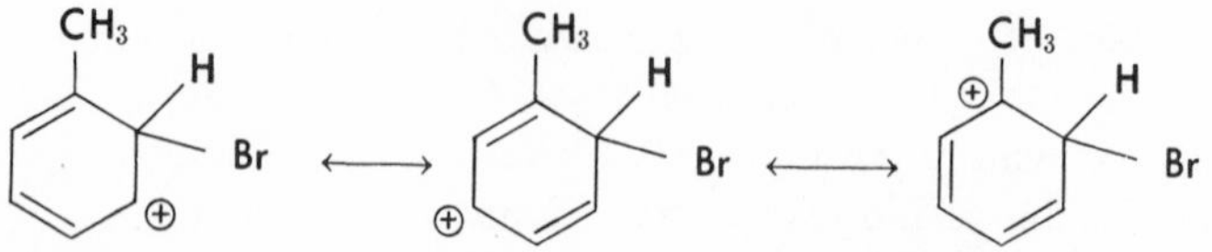

Such hindrance varies with the size of the attacking group. For chlorination of toluene in acetic acid the partial rate factors are: o_f^{Me}, 617; m_f^{Me}, 5; p_f^{Me}, 820. Thus here an ortho position is almost as reactive as the para, reflecting less hindrance in the transition state for ortho attack. At first sight it might be thought that this simply indicates that a chlorine atom is smaller than a bromine, but something more is involved. The partial rate factor for para chlorination is smaller than for bromination, although both charged intermediates look much the same. This is because the charged intermediates come *after* the transition states; in the transition state for aromatic substitution only part of the positive charge will yet be developed on the benzene ring. The smaller p_f^{Me} for chlorination suggests that it has an earlier transition state, in which less charge has yet developed. Thus there would in any case be less crowding in the transition state for ortho substitution, since the attacking chlorine molecule is further away.

It has so far not been possible to arrive at a good quantitative treatment for ortho substitution, because steric hindrance comes in as well as carbonium ion stability. However, the relative reactivities of meta and para positions of most compounds in most electrophilic substitutions can be successfully correlated with two sets of constants in a modified Hammett equation (cf. Chapter 4). Each substituent is assigned a substituent constant, called σ^+ (sigma plus) and each reaction is assigned a reaction constant, called ρ (rho). The substituent constant σ^+ measures the ability of a particular substituent, such as a *p*-methoxyl group, to stabilize the charged intermediate; σ^+ is related to σ of Chapter 4, but for substituents which have strong resonance interaction with a positive charge the value of σ^+ is different from that of σ. The reaction constant measures the susceptibility of the particular reaction to such stabilization (i.e., the extent to which the actual transition state resembles the charged intermediate). Then the partial rate factor for substitution para to a group R in a bromination, for instance, obeys the following relationship,

$$\log p_f^R = \sigma^+_{p\text{-R}} \cdot \rho_{\text{bromination}}$$

while a similar relationship governs reactivity at the meta positions

$$\log m_f^R = \sigma^+_{m\text{-R}} \cdot \rho_{\text{bromination}}$$

In Table 5–1 are listed values of σ^+ for some common substituents; in Table 5–2 are listed the ρ values for several typical aromatic substitutions.

TABLE 5–1 ▪

Substituent Constants for Aromatic Substitution

	σ^+ meta	σ^+ para
CH_3O-	0.047	−0.778
CH_3-	−0.066	−0.311
H−	0	0
Cl−	0.399	0.114
Br−	0.405	0.150
$Me_3\overset{+}{N}-$	0.359	0.408
HO_2C-	0.322	0.421
$-C\equiv N$	0.562	0.659
$-NO_2$	0.674	0.790

Negative σ^+ constants indicate activation, while a group with a positive σ^+ para, for instance, is deactivating to the para position. Thus going down the table, the methoxy group is strongly activating to the para position (and hence to the ortho position as well) while it is mildly deactivating to the meta position. A methoxy group stabilizes an adjacent positive charge by a resonance interaction in para substitution; the intermediate in meta substitution is slightly destabilized since its positive charge is near the positive end of a carbon-oxygen dipole.

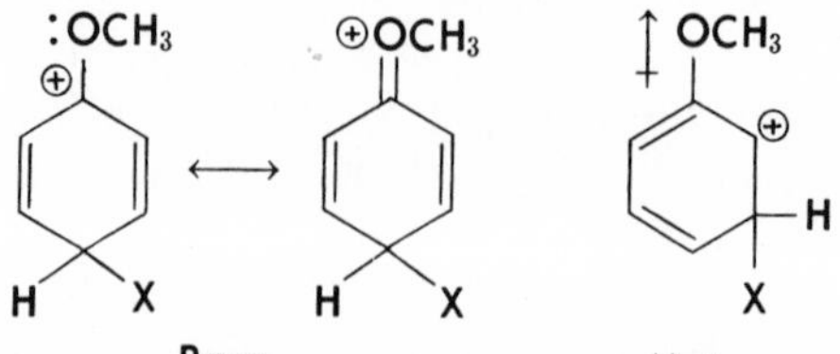

Para Meta

The σ^+ constants for methyl show that it is activating to both the meta and para positions, as was discussed above. The meta activation is very slight, and the para activation is less effective than for methoxy. Hydrogen is the standard, whose σ^+ constants are zero by definition. Chlorine and bromine deactivate both the meta and the para (and ortho) positions; since the meta position is most strongly deactivated, substitution occurs mostly ortho and para to the halogens. This can be understood in terms of an inductive effect which strongly withdraws electrons from the ring, and at the same time a possibility for some resonance stabilization of the intermediate in para (and ortho) substitution.

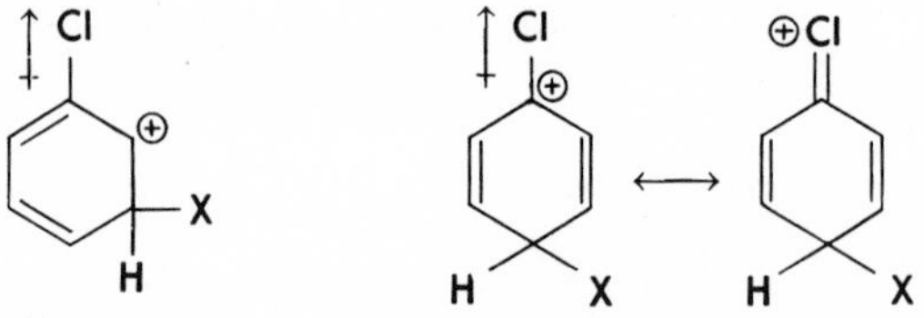

Meta Para

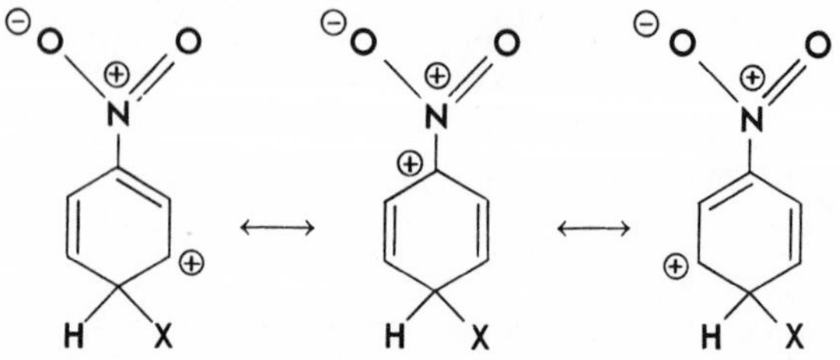

The last four groups listed deactivate the para position more than they deactivate the meta. They have either a positively charged atom or the positive end of a dipole attached to the ring, and they strongly destabilize the positively charged intermediates in aromatic substitution. The effect is larger for para (and ortho) substitution because one resonance form of the intermediate places the positive charge right next to the deactivating group, for instance the nitro group.

The values of ρ listed in Table 5–2 show how sensitive typical aromatic substitution reactions are to the effects of substituents. A large negative ρ indicates that the reaction is strongly affected, while the Friedel-Crafts alkylation with ethyl bromide has a small negative ρ, so it is rather insensitive to substituent effects. This is more apparent if we calculate the partial rate factors for bromination and for Friedel-Crafts ethylation para to the methoxy in anisole.

$$\log p_f^{\mathrm{OMe}} = \sigma^{+}_{p\text{-OMe}} \cdot \rho$$

For bromination,

$$\log p_f^{\mathrm{OMe}}\ (\text{bromination}) = (-\ 0.778)\ (-\ 12.1) = 9.414$$

thus

$$p_f^{\mathrm{OMe}}\ (\text{bromination}) = 2{,}600{,}000{,}000$$

The para position in anisole is over two billion times as reactive toward bromine as a simple benzene position would be. However, for ethylation,

$$\log p_f^{\mathrm{OMe}}\ (\text{ethylation}) = (-\ 0.778)\ (-\ 2.4) = 1.867$$

thus

$$p_f^{\mathrm{OMe}}\ (\text{ethylation}) = 74$$

Although there is still activation, the effect is much smaller. The transition state for Friedel-Crafts ethylation occurs very early, and not much charge has yet developed in the ring.

Similar calculations can be done for the other reactions, and

TABLE 5–2 ■

Reaction Constants for Aromatic Substitution

Reaction	ρ
Br_2 bromination, acetic acid, 25°C	−12.1
Cl_2 chlorination, acetic acid, 25°C	−10.0
Friedel-Crafts acetylation	− 9.1
HNO_3 nitration in nitromethane, 25°C	− 6.0
Friedel-Crafts ethylation (alkylation)	− 2.4

using other substituents. In general the agreement with experiment is very good, although it is not always perfect. However, a very important substitution reaction is missing from the table, nitration using nitric and sulfuric acids. This reaction does not fit the $\sigma^+ \rho$ relationship at all; the very interesting reason for this is discussed in Special Topic 5.

5–2 Nucleophilic Substitution

Chlorobenzene is rather inert under the usual conditions for nucleophilic displacement reactions. Apparently S_N2 displacement at an unsaturated carbon is quite an unfavorable process, while S_N1 reactions do not occur because of the instability of the resulting carbonium ion, a phenyl cation. Although nucleophilic substitution reactions of aromatic compounds are thus not as common as are electrophilic substitutions, they can occur under certain conditions.

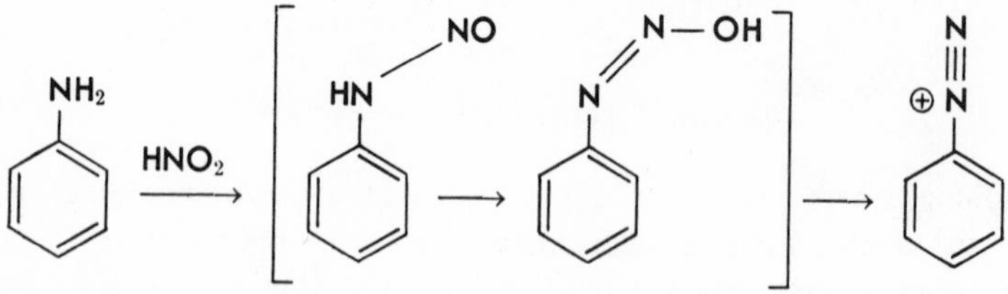

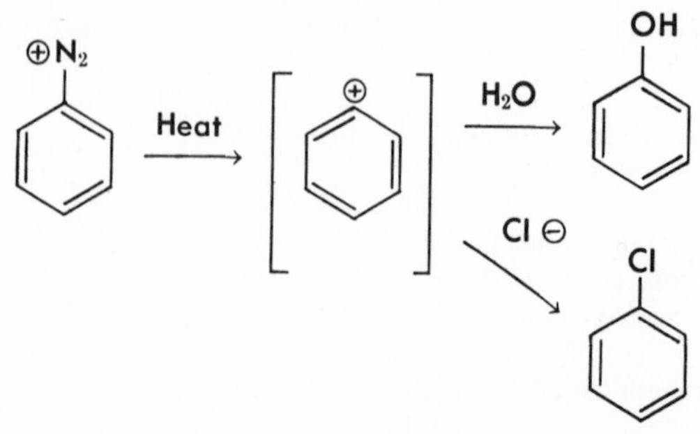

Aromatic diazonium compounds. When aniline is treated with nitrous acid it is converted to the benzenediazonium cation. This can be decomposed by heating in water, phenol being formed by a nucleophilic substitution reaction. In the presence of a high concentration of chloride ions the product is instead largely chlorobenzene, but the rate of decomposition of the diazonium salt is unaffected. This evidence shows that the rate-determining step does not involve the nucleophile, and supports an S_N1 mechanism. The very unstable phenyl cation is possible here only because the formation of N_2 is a very favorable process.

Addition-elimination on activated aromatic rings. Just as electrophilic substitution involves the formation of a positively charged intermediate, nucleophilic substitution may occur through a negatively charged intermediate. Such a process is facile only if the negative charge can be stabilized by appropriate substituents. Thus 2,4-dinitrochlorobenzene reacts with methoxide ion in methanol solution via such an intermediate anion; conjugation with the nitro groups makes this occur under conditions in which chlorobenzene is perfectly inert.

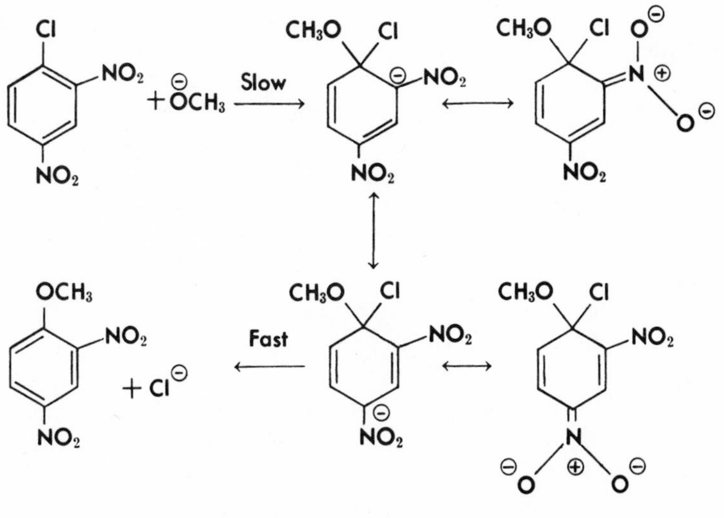

It is interesting that 2,4-dinitrofluorobenzene is much more reactive than is the chloro compound. As was seen in Chapter 3, fluoride ion is a poorer leaving group than is chloride ion, so the greater reactivity of the fluoro compound here is at first sight surprising. However, the rate-determining step in these substitutions is addition of the anion. Loss of the halide ion is a subsequent fast step, even with the slower fluoride ion. Addition to the ring is favored (by inductive and field effects) by the electron-withdrawing halogen, so the first step is faster with the fluoro compound.

A number of other groups may activate the ring in this way, e.g., carbonyl or sulfonyl groups. In addition, this type of mechanism is important in the chemistry of heterocyclic aromatic systems when the negative charge can be placed on electronegative atoms. For instance, 2-bromopyridine reacts with ammonia to afford 2-aminopyridine.

N Br + NH_3 ⟶ N Br NH_2 ⊖ ⟶ N NH_2

Substitution via benzyne. When *p*-bromoanisole is treated with potassium amide in liquid ammonia solution, it is converted to a mixture of *m*- and *p*-anisidines.

Br OCH_3 + KNH_2 $\xrightarrow{NH_3}$ NH_2 OCH_3 + NH_2 OCH_3

Furthermore, treatment under these conditions of chlorobenzene labeled with carbon-14 affords aniline in which the label is essentially equally distributed between the carbon bearing the amino group and a carbon ortho to it. These results are best rationalized by postulating that substitution proceeds through a benzyne, a cyclic acetylene.

Cl * H $\xrightarrow[NH_3]{NH_2^{\ominus}}$ Cl * ⊖ ⟶ * ⊖ NH_2 ⟶

*⊖ NH_2 + NH_2 * ⊖ ⟶ NH_2 * + NH_2 *

For simplicity the elimination of HCl is written as a two-step process, involving the carbanion, although this is not really certain.

Benzyne and substituted benzynes are now well known reactive intermediates, which are most conveniently generated in ways other than that above. For example, treatment of *o*-bromofluorobenzene with lithium in the presence of anthracene generates benzyne, which is trapped by the anthracene to yield the interesting compound triptycene.

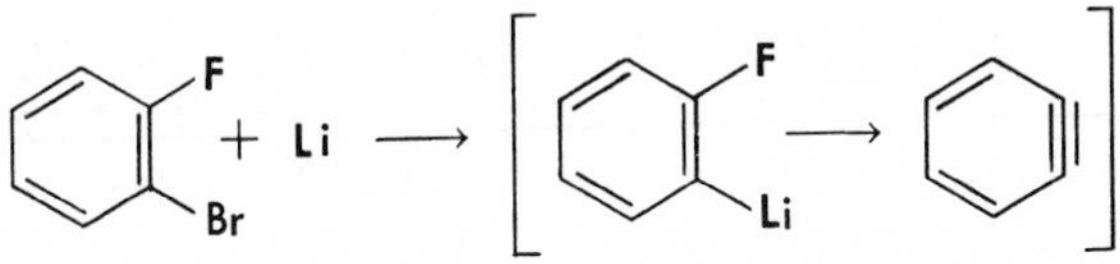

F Br + Li ⟶ [F Li ⟶]

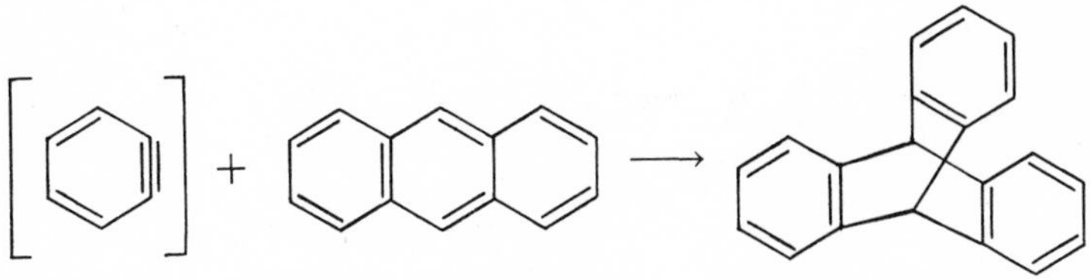

[ ] + ⟶

It is not surprising that benzyne is quite unstable and has only been detected as a reactive intermediate. Acetylenes ordinarily are linear, since they form σ bonds with *sp* hybrid orbitals and use the other *p* orbitals for π bonding. In benzyne this is not possible, and the new "π bond" which has been formed in the plane of the ring involves overlap of hybrid orbitals.

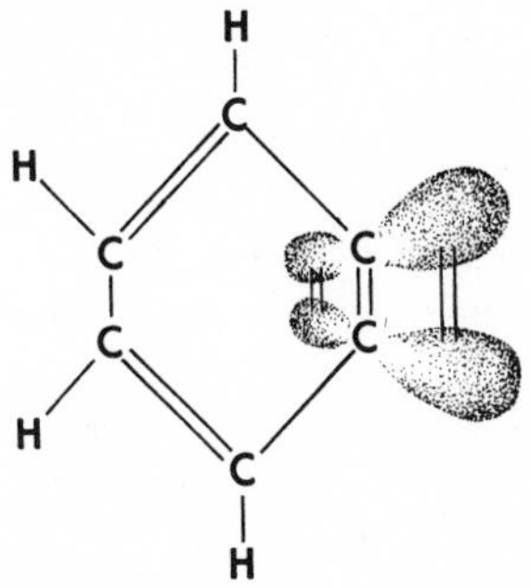

This strain decreases as the ring becomes larger; cycloöctyne is an isolable, if somewhat unstable, compound.

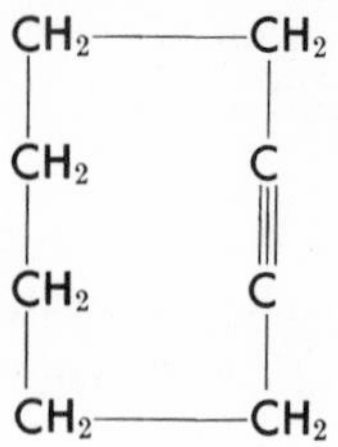

CH_2—CH_2
CH_2 C
CH_2 C
CH_2—CH_2

Direct displacement. With weaker bases than NH_2^-, direct substitution on aromatic halides may compete with benzyne formation. The hydrolysis of chlorobenzene with strong sodium hydroxide solution at a high temperature yields phenol; studies with radioactive carbon show that this largely, but not exclusively, involves benzyne as an intermediate.

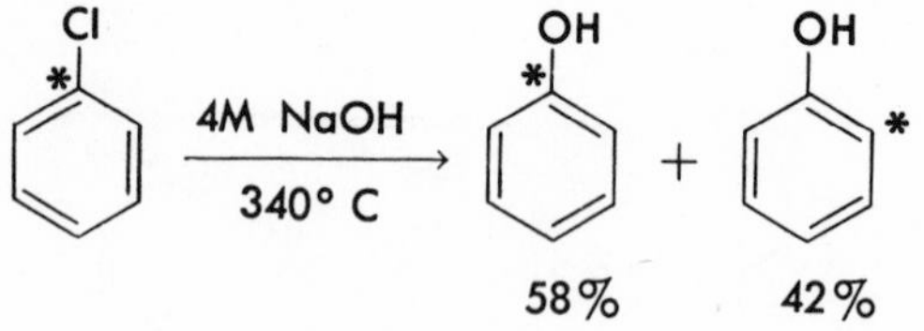

Cl — 4M NaOH, 340° C → OH + OH

58% 42%

The fact that more radioactivity is found in C–1 of the phenol shows that a direct substitution process occurs along with the benzyne mechanism. Apparently the direct displacement mecha-

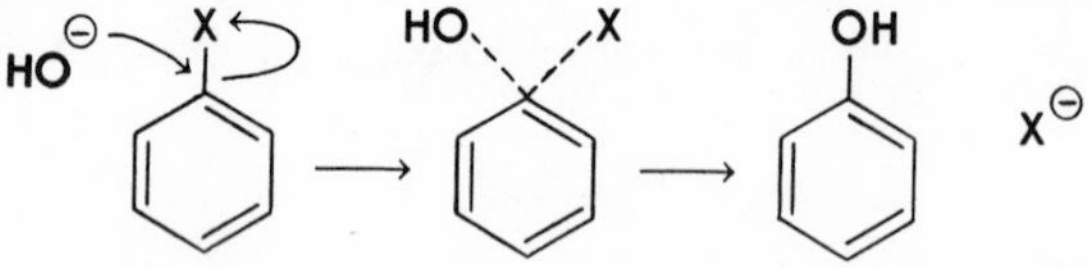

$HO^{\ominus}$ X → HO X → OH $X^{\ominus}$

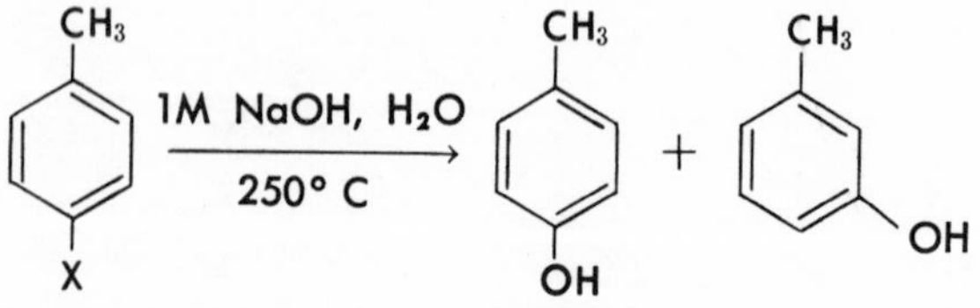

CH_3 X — 1M NaOH, H_2O, 250° C → CH_3 OH + CH_3 OH

X = Br 78.5% 21.5%

X = I ~97% <3%

nism resembles a simple S_N2 reaction; it becomes more favorable when the leaving group is better.

However, this S_N2-like reaction is obviously still quite difficult, requiring conditions much more vigorous than those used for displacements on simple alkyl halides.

General References

L. Stock, *Aromatic Substitution Reactions* (Prentice-Hall, Englewood Cliffs, New Jersey, 1968). A good modern introduction.

G. Olah, *Friedel-Crafts and Related Reactions* (Interscience Publishers, New York, 1963). A four-volume compendium which discusses in detail not only Friedel-Crafts acylation and alkylation, but also other aromatic substitution reactions, e.g., nitration and halogenation.

J. Hine, *Physical Organic Chemistry* (McGraw-Hill Book Co., 1962), Chapters 16 and 17. Discusses both electrophilic and nucleophilic substitutions.

L. M. Stock and H. C. Brown, "A Quantitative Treatment of Directive Effects in Aromatic Substitution," *Advances in Physical Organic Chemistry,* **1,** 35 (1963). A review of the evidence for the $\sigma^+\rho$ correlation.

J. F. Bunnett, "Mechanism and Reactivity in Aromatic Nucleophilic Substitution Reactions," *Quarterly Reviews* (*London*), **12,** 1 (1958). A good review of the S_N1, bimolecular, and benzyne mechanisms.

H. Heaney, "Benzyne and Related Intermediates," *Chemical Reviews,* **62,** 81 (1962).

G. Wittig, "Small Rings with Carbon-Carbon Triple Bonds," *Angewandte Chemie, International Edition,* **1,** 415 (1962).

K. Wiberg, "The Deuterium Isotope Effect," *Chemical Reviews,* **55,** 713 (1955). For more information on how isotope effects can be used to elucidate mechanisms.

L. Melander, *Isotope Effects on Reaction Rates* (Ronald Press Co., New York, 1960). Chapter 6 deals with aromatic substitution.

H. Zollinger, "Hydrogen Isotope Effects in Aromatic Substitution Reactions," in V. Gold, ed., *Advances in Physical Organic Chemistry,* Vol. 2 (Academic Press, New York, 1964), p. 163.

5

Special Topic

■ THE ROLE OF π COMPLEXES IN AROMATIC SUBSTITUTION

WHEN BENZENE IS ADDED to a mixture of HF and BF_3 a reaction occurs with formation of the highly colored σ complex.[1]

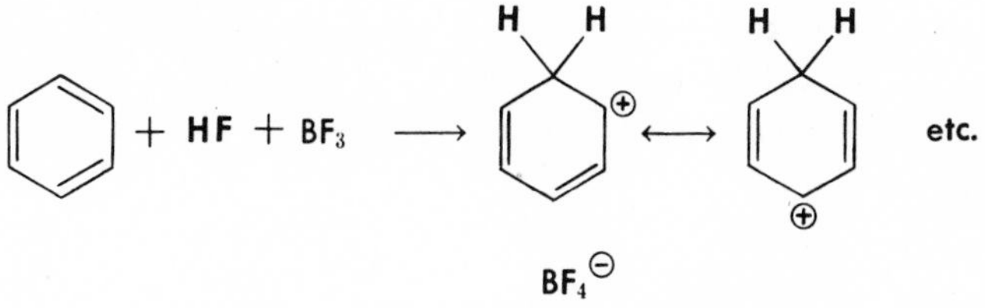

$$C_6H_6 + HF + BF_3 \longrightarrow C_6H_7^{\oplus}\,BF_4^{\ominus} \quad (\text{resonance forms} \longleftrightarrow \text{etc.})$$

A σ complex

A proton has been added to one of the benzene ring carbons, and in the resulting species the proton and the ring are thus "complexed" by the formation of a new σ bond. In the last chapter such σ

1. J. Hine, *Physical Organic Chemistry* (2nd ed., McGraw-Hill Book Co., New York, 1962), p. 348.

complexes were simply called "charged intermediates." Solutions of these complexes are electrically conducting, since the complexes are actually salts. Furthermore, when DF is used instead of HF the deuterium exchanges with the benzene protons, since in the σ complex the deuterium newly attached to the ring carbon is now equivalent to the proton already on that benzene carbon, so that either may be lost when the complex is decomposed. These properties, along with spectroscopic studies, support the structure assigned.

By contrast, a different sort of interaction occurs when HCl is dissolved in benzene.[1] Studies of the solubility of HCl show that it forms a 1:1 complex with benzene, but the solution has no color. Furthermore, it does not conduct electricity, so a salt has not been formed. Finally, when DCl is used no deuterium exchange with the ring protons occurs. This second type of complex is called a π complex. Apparently the proton of HCl has a weak interaction with the π electrons of the benzene ring without being transferred away from the chlorine atom.

(benzene) + HCl ⟶ (benzene) ⟶ H—Cl

A π complex

The bonding in π complexes can be described in either molecular orbital or valence bond terms. In a molecular orbital picture the hydrogen 1s orbital has a little overlap with the two carbon p orbitals of a π bond, while retaining most of its overlap with the chlorine orbital.

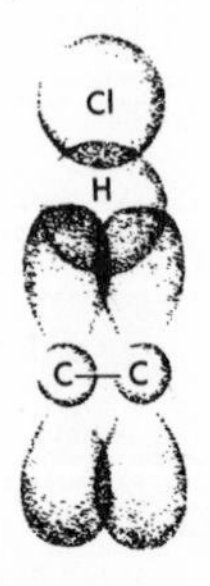

In valence bond terms the π complex is a hybrid of three principal resonance forms; the neutral one is most stable and contributes most to the structure.

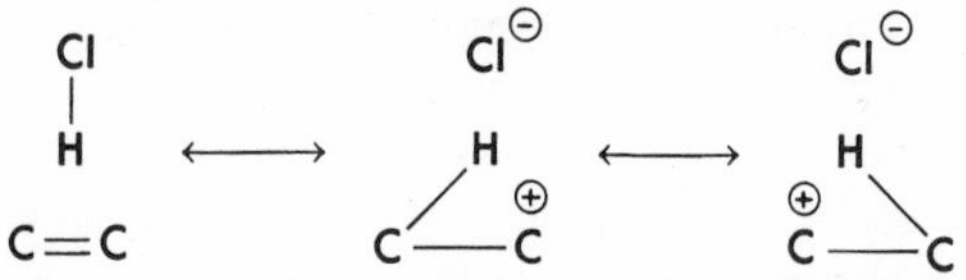

The valence bond picture shows that there is some polar character to these complexes, and they are often called "charge-transfer complexes."[2]

A variety of electrophiles can give such π complexes (charge-transfer complexes) with aromatic compounds. Silver ion forms a complex with benzene and other aromatic rings (and also with simple olefins), halogens such as Cl_2 form π complexes with aromatic compounds under mild conditions in which substitution does not occur (e.g., low temperature), and polynitro compounds such as picric acid form crystalline complexes with many aromatic hydrocarbons.[2]

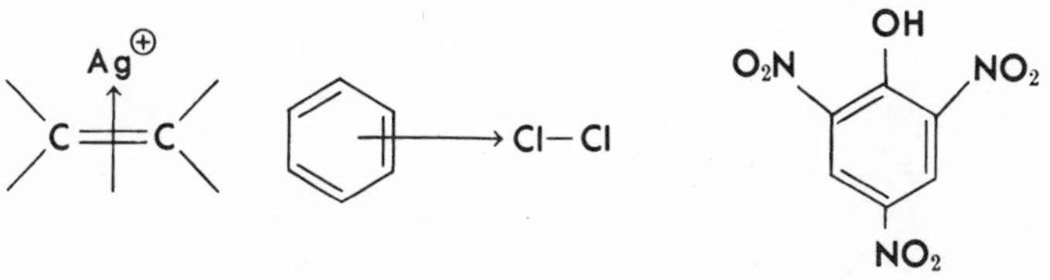

Picric acid

The exact structure of a π complex depends on the nature of the donor (the molecule, such as benzene, which partially donates electrons and acquires a little positive charge) and the acceptor. An x-ray diffraction study of the complex of benzene with silver perchlorate shows that the silver ion lies above one particular carbon-carbon bond,[2] while in the benzene-bromine complex the bromine lies above the center of the benzene ring[2] (the bromine

2. For a good general discussion, cf. L. Ferguson, *The Modern Structural Theory of Organic Chemistry* (Prentice-Hall, Englewood Cliffs, New Jersey, 1963), p. 103.

molecule lies on the axis of the ring, the nearest bromine atom interacting equally with all six *p* orbitals). In a complex of picric acid with an aromatic hydrocarbon the two lie face-to-face.[2] Overlap of the two π systems is stabilizing since there is then some charge transfer from the donor molecule to the picric acid (with three electron-attracting nitro groups).

Since electrophiles can form π complexes with aromatic compounds, and many electrophiles also carry out substitution on benzene rings, one might wonder whether π complexes play a role in substitution reactions. For instance, the following scheme can be pictured for the bromination of benzene.

+ Br_2 ⇌ → Br—Br → $Br^{\ominus}$ H Br ⊕ →

π Complex σ Complex

Br → H—Br → Br + HBr

π Complex

This sequence is probably correct. However, the first step, formation of a π complex, is rapid and reversible and the transition state occurs during the conversion of the π complex to the σ complex. Thus in discussing the effects of substituents on aromatic substitution reactions there is no need to take account of the intermediate formation of a π complex, for the transition state is what matters and it strongly resembles the σ complex. The strongest evidence for this is found in Table 5–3, in which the relative rates of bromination in acetic acid solution are compared with the equilibrium constants for σ-complex formation (HBF_4) and π-complex formation (HCl) for a series of methylbenzenes.[1]

It is quite apparent from this table that both the rate of halogenation and the stability of a σ complex are strongly increased by methyl substituents, while the stability of a π complex is

TABLE 5–3 ▪

Rates of Bromination of some Methylbenzenes Compared with the Equilibrium Constants for σ Complex and π Complex Formations

Benzene Derivative	Relative Bromination Rate	Relative σ Complex Stability	Relative π Complex Stability
Benzene	1	1	1
Toluene	605	7	1.51
1,4-Dimethylbenzene	2500	11	1.65
1,2-Dimethylbenzene	5300	12	1.85
1,3-Dimethylbenzene	514,000	290	2.06
1,2,4-Trimethylbenzene	1,520,000	700	2.23
1,2,3-Trimethylbenzene	1,670,000	770	2.40
1,2,3,4-Tetramethylbenzene	11,000,000	4400	2.68
1,3,5-Trimethylbenzene	189,000,000	145,000	2.60
1,2,3,5-Tetramethylbenzene	420,000,000	178,000	2.74
Pentamethylbenzene	810,000,000	322,000	—

affected much less. This difference is to be expected. In a π complex only a little charge is transferred to the ring, since the principal resonance form is the neutral one. Furthermore, this charge is placed in π orbitals delocalized over all six benzene carbons, and it can be stabilized more or less equally by methyl groups in any position. Thus there is little difference in π basicity among the three isomeric xylenes (dimethylbenzenes), or among the three trimethyl isomers or the two tetramethyls.

On the other hand, the relative stabilities of σ complexes are strongly affected by both the number and the position of methyl substituents. *m*-Xylene (1,3-dimethylbenzene) is almost 30 times as basic as the other xylenes, since both methyls are on charged carbons.

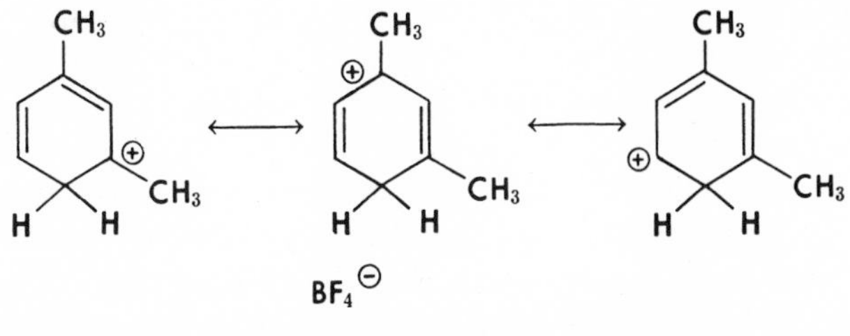

CH_3 CH_3 CH_3 H H ⟷ H H ⟷ H H $BF_4^{\ominus}$

As expected, mesitylene (1,3,5-trimethylbenzene) is even more basic, since the third charged carbon now bears a methyl group, although the 500-fold effect of this last group seems rather large. This trimethylbenzene is even more basic than is 1,2,3,4-tetramethylbenzene; in the σ complex from the latter only two of the positive carbons carry methyl groups. Relative reactivities toward bromine also follow this order, although the π complex stabilities are reversed.

Table 5–3 clearly indicates that the transition state for halogenation resembles the σ complex, and not the π complex. In fact, halogenation is even more sensitive to substituent effects than σ complexing is. This surprising result may reflect subtle differences in the reaction conditions, since it is not reasonable that the transition state for bromination could have more positive charge on the benzene ring than is found in a σ complex.

As we saw in Chapter 5 (Table 5–2), of all the common substitution reactions bromination is the most sensitive to substituent effects. However, the other reactions listed in Table 5–2 also show relatively large increases in rate for methylated benzenes; consequently their transition states also resemble the σ complex more closely than the π complex. This was also implied in our use of partial rate factors. The concept of a partial rate factor involves the idea that in the transition state the electrophilic reagent is attacking a *particular* carbon atom of the benzene ring, on the way to a σ complex. If the transition state were instead on the way to a π complex this treatment would not have worked. In fact, it fails for nitration by nitronium ion. As is discussed below, this failure seems to be strong evidence that in this case the rate-determining step is formation of the π complex.

The nitration of a series of aromatic hydrocarbons has been studied,[3] using nitronium fluoborate in tetramethylene sulfone as solvent. Relative to benzene, toluene has a reactivity of 1.67, *m*-xylene a reactivity of 1.65, and mesitylene a reactivity of only 2.71. These very small effects of the methyl groups are very similar to their effects on the stability of π complexes, suggesting that the transition state in this case looks like a π complex, but these data alone are not enough to establish the point. The rate-determining step might be formation of a σ complex, but the transition state

3. G. Olah, S. Kuhn, and S. Flood, "Aromatic Substitution. VIII," *Journal of the American Chemical Society,* **83,** 4571 (1961); see also C. Ritchie and H. Win, "Some Nitronium Tetrafluoroborate Nitrations," *Journal of Organic Chemistry,* **29,** 3093 (1964).

might occur so early that very little charge had yet developed on the benzene ring. However, an attempt to calculate partial rate factors indicates that something is wrong with this picture. Nitration of toluene leads to only 2.8% meta substitution, with 65.4% ortho and 31.8% para. Thus the partial rate factor for the meta position of toluene, m_f^{Me}, can be calculated from these data:

$$m_f^{Me} = (1.67)\ (0.028)\ (3) = 0.14$$

The factor of 3 was included to take account of the fact that benzene has six equivalent positions while toluene has only two meta positions; the partial rate factor is the reactivity of a meta position compared to a single benzene position

The conclusion of this calculation is that a methyl group deactivates the meta position, reducing its reactivity to 14% of that of a benzene carbon. This is completely contradictory to all evidence from other reactions. As was discussed in Chapter 5, σ^+ for *m*-methyl is -0.066, indicating that it is an activating group, although a weak one. It seems that the partial rate factor treatment is not valid for this nitration.[3]

The situation becomes clear if we consider the possibility that the rate-determining step in this case is formation of a π complex.

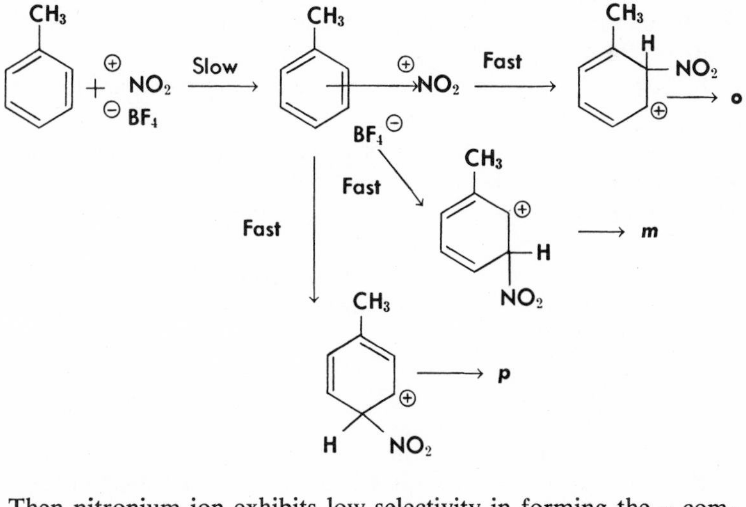

Then nitronium ion exhibits low selectivity in forming the π complex, since π complexes of toluene are only slightly stronger than

those of benzene. On the other hand, when the π complex collapses it may go to three different σ complexes, leading respectively to ortho, meta, and para substitution, and it tends to collapse mostly to the better σ complexes. Since in the rate-determining step it is not yet clear which carbon will finally be attacked, it is not surprising that the partial rate-factor treatment does not work.

These conclusions have been attacked by Tolgyesi[4] on the basis of work which seemed to show that reaction occurs while the reagents were being mixed. If the nitronium fluoborate reacted before the solution became homogeneous, then the reaction would seem to be nonselective among substrates for purely mechanical reasons (scavenging of all aromatic molecules in the vicinity of a drop of nitrating solution). However, Olah has since shown[5] that this does not occur, so at the present time the original interpretation—rate-determining formation of a π complex—seems the most likely.

Similar studies have been conducted with nitronium salts in other solvents,[6] and the results are the same. The same selectivity is also observed[6] in nitration with nitric-sulfuric acid mixtures, in which the active species is nitronium ion. However, nitration with nitric acid in nitromethane and other organic solvents probably does not involve nitronium ion,[6] since the toluene/benzene reactivity is 26, a high relative reactivity consistent with rate-determining σ complex formation. The 3.1% meta substitution which is observed leads to a m_f^{Me} of 2.4; this activation is as expected for rate-determining σ complex formation.

It is interesting that the rate of nitration of toluene with nitric acid in nitromethane is independent of the toluene concentration.[5] This and other kinetic evidence had been taken to mean that the

4. W. Tolgyesi, "Relative Reactivity of Toluene-Benzene in Nitronium Tetrafluoroborate Nitration," *Canadian Journal of Chemistry,* **43,** 343 (1965).
5. G. Olah and N. Overchuk, "Remarks on the Nitronium Salt Nitration of Toluene and Benzene," *Canadian Journal of Chemistry,* **43,** 3279 (1965).
6. G. Olah, S. Kuhn, S. Flood, and J. Evans, "Aromatic Substitution. XIII," *Journal of the American Chemical Society,* **84,** 3687 (1962).
7. C. Ingold, *Structure and Mechanism in Organic Chemistry* (Cornell University Press, Ithaca, New York, 1953), p. 275.

rate-determining step was formation of nitronium ion,[7] but as we have seen nitronium ion is probably not involved since in other solvents it gives toluene/benzene selectivities of 1.7 or so, not the observed 26. The true nitrating agent in this system (N_2O_5?) has not yet been identified.

It would be expected that π complex formation could be rate determining whenever the attacking reagent is so electrophilic that the π complex cannot dissociate once it is formed. This has also been found[8,9] to be true in some Friedel-Crafts alkylations. Thus when toluene is alkylated with benzyl chloride and aluminum chloride in nitromethane only 4.5% meta substitution occurs, indicating high position selectivity. However, toluene is only 3.2 times as reactive as benzene, indicating low substrate selectivity; this combination of facts, which would lead to a m_f^{Me} of 0.43, shows again that the rate-determining step is π complex formation, and that the product distribution is determined later.

$C_6H_5CH_2Cl$ + (CH_3) + $AlCl_3$ —Rate-Determining→ (CH_3, $\oplus CH_2C_6H_5$) —Fast→

(CH_3, ⊕, H, $CH_2C_6H_5$) + (CH_3, ⊕, $CH_2C_6H_5$, H) + (CH_3, ⊕, H, $CH_2C_6H_5$) —Fast→ Products

In such cases the partial rate-factor treatment, and the $\sigma^+ \rho$ relationship which was applied to it, are not valid. However, for most aromatic substitutions π complex formation is just a very probable first reversible step, and σ complex formation is rate-determining.

8. G. Olah, S. Kuhn and S. Flood, "Aromatic Substitution. X," *Journal of the American Chemical Society,* **84,** 1688 (1962).
9. G. Olah and N. Overchuk, "Aromatic Substitution. XXV," *Journal of the American Chemical Society,* **87,** 5786 (1965).

6

REACTIONS OF CARBONYL COMPOUNDS

A CARBONYL GROUP conveys two characteristic types of reactivity to a molecule. Through enol or enolate ion formation a carbonyl group activates adjacent C—H bonds toward substitution, while there are also a number of characteristic addition and substitution reactions which occur at the carbonyl group itself. Activation of an adjacent C—H is characteristic of all types of carbonyl compounds—ketones, aldehydes, esters, amides, etc.—and it will be considered first. However, since ketones and aldehydes undergo reactions at the carbonyl group which are rather different from those of esters and other carboxyl derivatives, these will be considered separately in the second and third parts of this chapter.

6–1 Enols and Enolates

As was discussed in Chapter 2, the bromination of acetone occurs by enolization, followed by bromination of the enol. The kinetic expression, when the reaction is conducted in an aqueous acetic acid/sodium acetate buffer, indicates that both acids and bases catalyze the process. With an acid it seems clear that the enol is formed, but for the process catalyzed by OH^-, for instance, the reactive intermediate is probably the enolate ion instead.

$$CH_3-\overset{\overset{O}{\|}}{C}-CH_3 + OH^{\ominus} \longrightarrow CH_3-\overset{\overset{O^{\ominus}}{|}}{C}{=}CH_2 \xrightarrow{Br_2} CH_3\overset{\overset{O}{\|}}{C}CH_2Br + Br^{\ominus}$$

The distinction is real, but it is difficult to make experimentally, and we shall often not distinguish between enols and enolate ions in discussing base-catalyzed processes.

Quite a complete study has been done on phenyl *sec*-butyl ketone (I), which is optically active. The rate of acid-catalyzed iodination is identical with the rate of racemization under the same conditions, and the rates of base-catatyzed bromination, deuteration, and racemization have also been found to be the same. All involve enol (or enolate ion) formation as the rate-determining step.

$$\underset{\text{I}}{(CH_3CH_2)(CH_3)(H)C-\overset{\overset{O}{\|}}{C}C_6H_5} \xrightarrow{\text{Slow}}$$

$$(CH_3CH_2)(CH_3)C{=}C(OH)C_6H_5 \xrightarrow[I_2]{Br_2} (CH_3CH_2)(CH_3)CX-\overset{\overset{O}{\|}}{C}C_6H_5$$

$$\downarrow D_2O$$

$$(CH_3-CH_2)(CH_3)(D)C-\overset{\overset{O}{\|}}{C}C_6H_5 + (CH_3CH_2)(D)(CH_3)C-\overset{\overset{O}{\|}}{C}C_6H_5$$

Acetone contains, at equilibrium, only 0.00025% enol. As Table 6–1 shows, however, other ketones may be much more enolic. In acetylacetone the enol is stabilized by an additional resonance form.

TABLE 6–1 ▪

Per Cent Enol at Equilibrium (in the Pure Liquid)

Compound	% Enol
Acetone	0.00025
Cyclohexanone	0.020
Acetylacetone	80
Acetoacetic ester	7.5
Biacetyl	0.0056
1,2-Cyclohexanedione	100

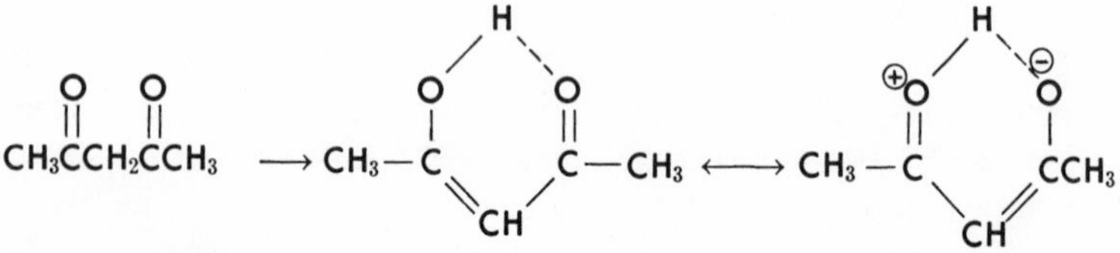

$$CH_3\overset{O}{\overset{\|}{C}}CH_2\overset{O}{\overset{\|}{C}}CH_3 \longrightarrow CH_3-C(-O-H\cdots O=)C-CH_3 \;(\text{via } CH) \longleftrarrow\!\!\longrightarrow CH_3-C(=\overset{\oplus}{O}-H\cdots\overset{\ominus}{O}-)CCH_3 \;(\text{via } CH)$$

This stabilization becomes less effective in the enol of acetoacetic ester, for the ester group has less electron-withdrawing ability.

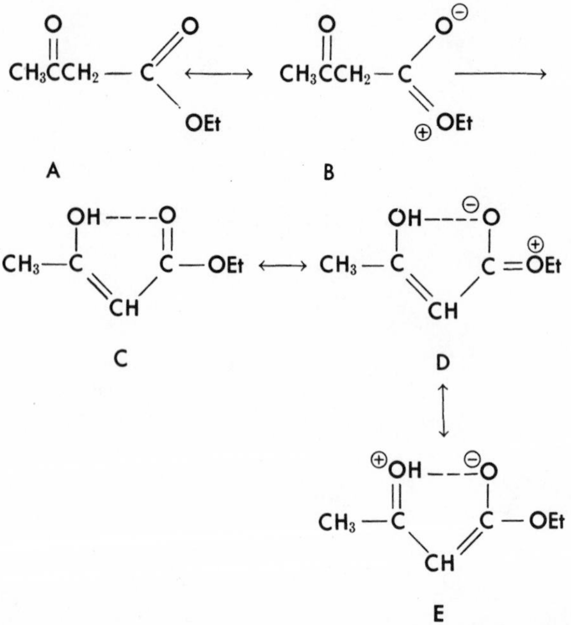

$$CH_3\overset{O}{\overset{\|}{C}}CH_2-C(=O)-OEt \longleftrightarrow CH_3\overset{O}{\overset{\|}{C}}CH_2-C(-O^{\ominus})=\overset{\oplus}{O}Et \longrightarrow$$

A B

$$CH_3-C(-OH\cdots O=)C-OEt \;(\text{via } CH) \longleftrightarrow CH_3-C(-OH\cdots{}^{\ominus}O-)C=\overset{\oplus}{O}Et \;(\text{via } CH)$$

C D

$$\updownarrow$$

$$CH_3-C(={}^{\oplus}OH\cdots{}^{\ominus}O-)C-OEt \;(\text{via } CH)$$

E

This effect can be described simply by saying that since the ethoxy group already feeds electrons into the carbonyl, the latter has less tendency to accept electrons from the enol group. Alternatively, in resonance terms we note that the starting ketone has two resonance forms for the ester group, A and B, while the product enol has only forms C, D, and E. Forms C and D are the ordinary ester resonance forms without any special interaction with the enol, and form E is the only one in which the enol interacts with the ester carbonyl group. This special interaction thus plays only a minor role in the structure of the enol; consequently the enol is less stabilized than it was for acetylacetone.

Comparing acetone with cyclohexanone, it is seen that the cyclic ketone is much more enolic. This is true chiefly because the enolization of acetone involves loss of freedom of rotation when the carbon-carbon single bond becomes a rigid double bond. Cyclohexanone is already rigid, so rotational freedom is not lost on enolization.

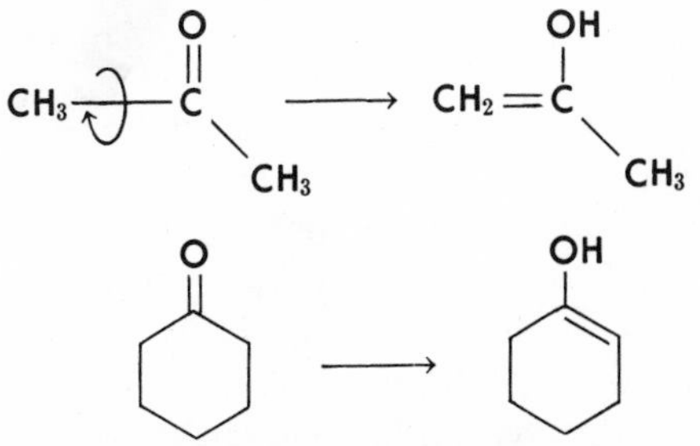

This is also partly involved in the difference between biacetyl and 1,2-cyclohexanedione, but another big factor comes from the conformations of the two starting materials. In biacetyl the two carbonyl groups are oriented so as to minimize dipole-dipole repulsion, but in the cyclic compound they are held so that this repulsion is quite strong. Enolization helps to relieve this dipole opposition; therefore cyclohexanedione is completely enolic even though the enol oxygen is not conjugated with the second carbonyl group.

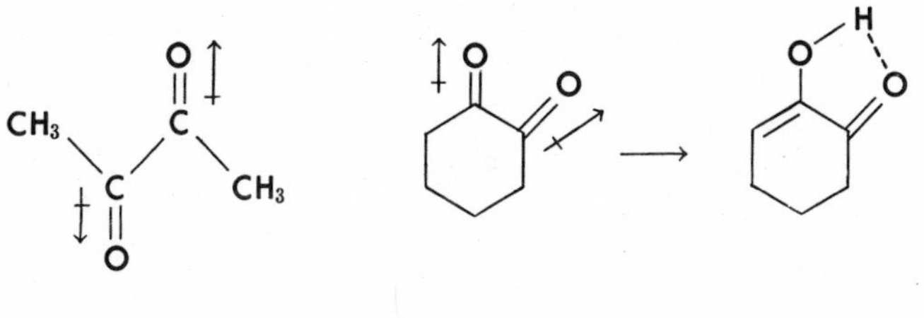

We have not referred to the effect of hydrogen bonding on the stabilities of these enols, although for the dicarbonyl compounds the enol was in all cases written with an internal hydrogen bond. This is undoubtedly a stabilizing factor as well, and is another one of the reasons that diketones are more enolic than monoketones. It is not required for enolization, however, for 1,3-cyclohexanedione derivatives are almost completely enolized even though they cannot form internal hydrogen bonds; in such cases the usual factors of conjugation in the enol and restricted rotation in the ketone are enough.

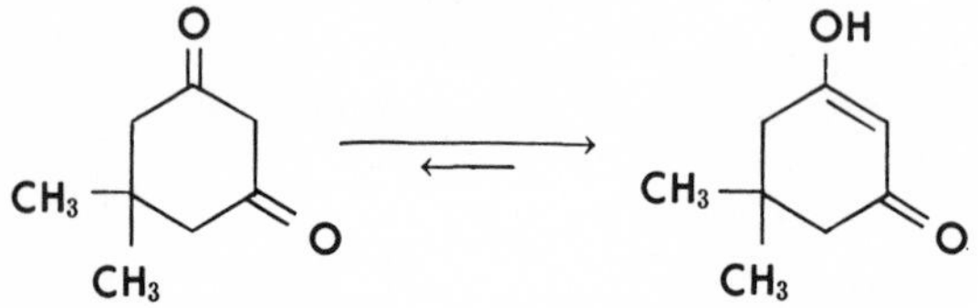

Since, as we have seen, the rate-determining step for many reactions involving enols is the enolization itself, relative rates of enolization are of more interest than the equilibrium values just discussed. The best data available are for the rates of enolate ion formation, so they will be of use chiefly in considering base-catalyzed processes. In Table 6–2 there are listed, for some representative compounds, both the rate constant for removal of a proton by water and the equilibrium constant K_a for this reaction.

TABLE 6–2 ▪ $R{-}H \underset{k_2}{\overset{k_1}{\rightleftarrows}} R^{\ominus} + H_3O^{\oplus} \qquad K_a = \frac{k_1}{k_2};\ 52°C$

	Compound	k_1 (per second)	K_a
(1)	CH_3COCH_3	4.7×10^{-10}	10^{-20}
(2)	CH_3COCH_2Cl	5.5×10^{-8}	3×10^{-17}
(3)	$CH_3COCHCl_2$	7.3×10^{-7}	10^{-15}
(4)	$CH_3COCH_2COCH_3$	1.7×10^{-2}	1.0×10^{-9}
(5)	$CH_3COCH_2CO_2C_2H_5$	1.2×10^{-3}	2.1×10^{-11}
(6)	$CH_3CO_2C_2H_5$	—	3×10^{-25}
(7)	$CH_2(CO_2C_2H_5)_2$	2.5×10^{-5}	5×10^{-14}
(8)	$CH_2(CHO)_2$	—	1×10^{-5}
(9)	CH_3CONH_2	3×10^{-14}	10^{-25}
(10)	CH_3CN	7×10^{-14}	10^{-25}
(11)	$CH_2(CN)_2$	1.5×10^{-2}	6.5×10^{-12}
(12)	$CH_3SO_2CH_3$	3×10^{-12}	10^{-23}
(13)	CH_3NO_2	4.3×10^{-8}	6.1×10^{-11}

Of course when a stronger base than water is used the rates will be faster, but the relationship of rates for different compounds in general will not be affected.

Examining the first three entries in the table, we see that the rate of enolate ion formation increases as the equilibrium stability of the enolate increases. The greater acidity of chlorinated acetones is due to the inductive effect of the chlorines; this effect carries over to the *rate* of ionization as well, since the transition state strongly resembles the enolate ion.

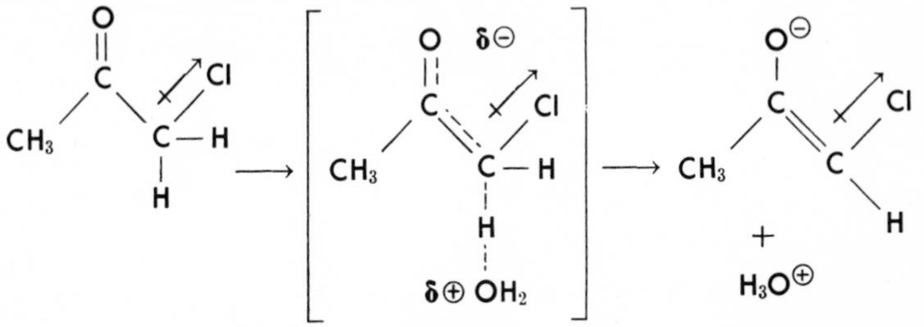

The relatively large acidity of acetylacetone (4), $pK_a = 9$, is also reflected in a high rate of ionization, while the related acetoacetic ester (5) is less acidic and less rapid in ionizing. Resonance forms similar to those written for the enols explain both the activating effect of a second carbonyl group and the lower effectiveness of this carbonyl if it is part of an ester group.

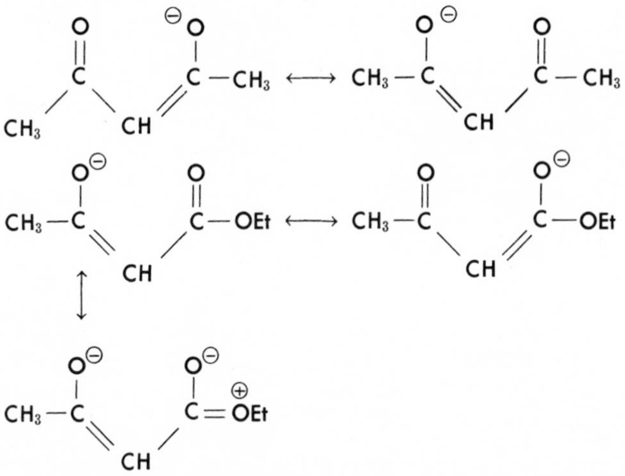

This weaker activation by an ester group is reflected as well in the fact that ethyl acetate (6) is a weaker acid than is acetone (1), and malonic ester (7) is both less acidic and slower to ionize than related keto compounds (4,5). In a sense, ester resonance is lost in the enolate ion.

An aldehyde group is even more activating than a ketonic carbonyl, as the high acidity of malondialdehyde (8), $pK_a = 5$, shows. This is to be expected since a methyl group is electron-feeding relative to a hydrogen; consequently the ketone carbonyl is a little less electronegative than an aldehyde carbonyl.

$$-\overset{\overset{\displaystyle O}{\|}}{C}-H \quad \text{versus} \quad -\overset{\overset{\displaystyle O}{\|}}{C} \longleftarrow CH_3$$

The carboxamide group is somewhat less activating (9) than an ester group, since electron donation into the carbonyl by the amino group is quite pronounced. A cyano group is also activating (10,11) and is somewhat more effective than an ester group.

$$^{\ominus}CH(C{\equiv}N)_2 \longleftrightarrow CH(C{\equiv}N)(C{=}N^{\ominus}) \longleftrightarrow CH(C{\equiv}N)(C{=}N^{\ominus})$$

The last two entries in the table show that both sulfonyl (12)

$$\overset{\ominus}{C}H_2-\overset{\oplus}{N}(=O)-O^{\ominus} \longleftrightarrow CH_2=\overset{\oplus}{N}(-O^{\ominus})-O^{\ominus}$$

$$\overset{\ominus}{C}H_2-\overset{(++)}{S}(-O^{\ominus})_2-CH_3 \longleftrightarrow CH_2=\overset{\oplus}{S}(-O^{\ominus})_2-CH_3$$

and nitro (13) groups are also acidifying. For the nitro group an ordinary resonance form can be drawn, but for sulfonyl it is necessary to place more than eight electrons around sulfur. This is possible since sulfur, a second-row element, has 3*d* orbitals available of relatively low energy and can accommodate more than eight electrons: two in its 3*s* orbital, six in its three 3*p* orbitals, and two or more distributed among its five 3*d* orbitals. Actually in the sulfone group itself there is already some double bonding involving the use of these *d* orbitals, so the doubly charged sulfur we have written is not correct.

We have repeatedly emphasized the parallel between rate of ionization and stability of the product enolate ion, pointing out that the transition state resembles the product. This is not always the case, however. For instance, nitromethane (13) is more acidic, $pK_a = 10.2$, than is malonic ester (7), $pK_a = 13.3$, but nitromethane ionizes more slowly. However, among more closely related compounds the parallel is quite good, and the type of structural arguments we have been using can be employed in predicting either relative rates or relative acidities.

This is another example of the Brønsted relationship described in Chapter 2. In order for the rate constants of ionization k_1 to run strictly parallel to equilibrium acidity constants K_a, rate constants of reprotonation k_2 would have to be uniform. For acetone (1) k_2 $(= k_1/K_a)$ is 4.7×10^{10} liters/mole/sec, but it is only 5.8×10^7 liters/mole/sec for acetoacetic ester (5). Thus the full effect of the equilibrium acidity is not seen in the rate. The theoretical upper limit for the bimolecular rate constant of a diffusion controlled process, in which reaction occurs at every collision, is about 5×10^{10} liters/mole/sec for two ions in water. Thus acetone enolate is protonated by H_3O^+ on almost every encounter. By contrast, the anion of nitromethane (13) has k_2 of only 6.8×10^2 liters/mole/sec. This is of course the rate constant for protonation on the carbon; protonation at oxygen is much faster.

We have not mentioned one special structural effect which can be important: the change in geometry which accompanies enolization. For instance, it has been found that nitrocyclopropane is a very weak acid compared with ordinary nitro compounds. This reflects the fact that in the anion the ring carbon must become sp^2 hybridized so that the remaining *p* orbital can be conjugated with the nitro group. The optimal angles for an sp^2 carbon are 120°, while for an sp^3 carbon they are 109°28′. Since the actual cyclopropane angle is 60°, the angle strain is greater for the anion than for the starting nitro compound, so I-strain suppresses ionization (cf. similar effects discussed in Chapter 3).

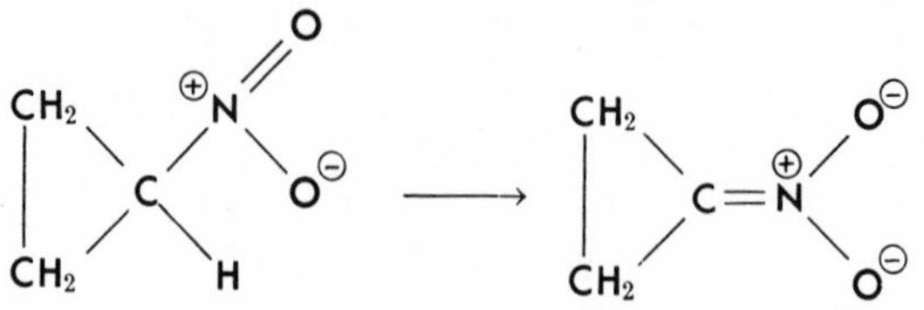

$$\text{cyclo-}(CH_2)_2C(H)-N^{\oplus}(=O)O^{\ominus} \longrightarrow \text{cyclo-}(CH_2)_2C=N^{\oplus}(O^{\ominus})_2$$

Strain = 109°28′ − 60°C = 49°28′ Strain = 120° − 60° = 60°

A different example is found in the fact that compound I is not particularly acidic, although it is a β-diketone. Here the geometrical problem is that all four groups on a double bond must be in the same plane, but this would not be possible for the enolate of I (this is best seen with molecular models).

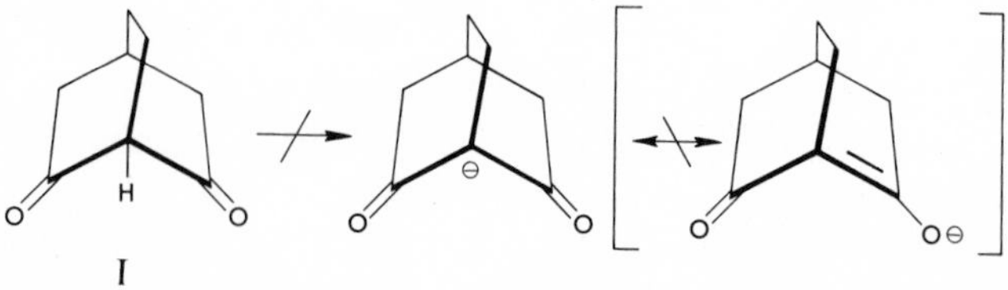

I

Enolate ions can be alkylated. Thus the malonic ester anion reacts with methyl iodide to form a new carbon-carbon bond, and this type of reaction is of considerable use in synthesis.

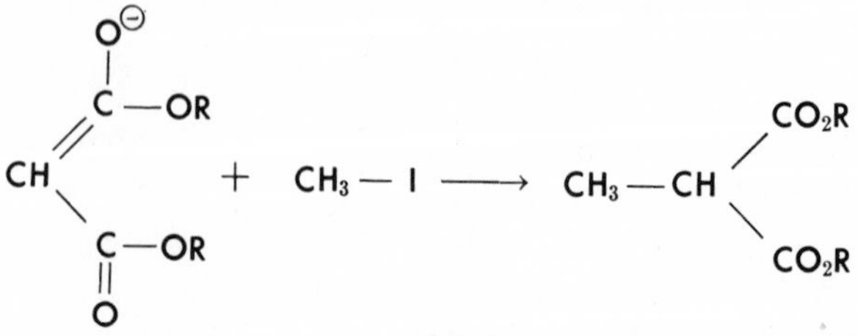

$$CH(=C(O^{\ominus})-OR)-C(=O)-OR + CH_3-I \longrightarrow CH_3-CH(CO_2R)_2$$

The reaction is simply an S_N2 displacement on methyl iodide by the enolate ion. It will be noted that the enolate ion reacts at its carbon, although most of the negative charge is distributed over the two oxygen atoms. This mode of reaction occurs because in this case the transition state for reaction is fairly far along toward product, and reaction at carbon leads to the more stable product and hence to the better transition state. However, when enolate ions are allowed to react with strong acids they protonate in a very exothermic reaction, and the transition state comes very early. Accordingly the fastest protonation occurs on oxygen to yield the enol, and only more slowly does this isomerize to the stable keto form. Some very exothermic alkylations have also been found to lead to attack on oxygen, forming enol ethers. The factors which lead to one or the other mode of attack are still under study.

One of the most important synthetic reactions of enolate ions is

$$\mathrm{>C{=}C(O^{\ominus})<} \xrightarrow{\mathrm{RX}} \mathrm{>C{=}C(OR)<} \quad \textbf{attack at center with most negative charge}$$

$$\mathrm{>C{=}C(O^{\ominus})<} \xrightarrow{\mathrm{RX}} \mathrm{R{-}\overset{|}{\underset{|}{C}}{-}\overset{O}{\overset{\|}{C}}{-}} \quad \textbf{attack to form more stable product}$$

their addition to carbonyl groups, as in the aldol condensation. This will be discussed at the end of the next section.

6–2 Addition to Ketones and Aldehydes

Reaction at the carbonyl group is the other characteristic process for ketones and aldehydes. These reactions may be classed as (1) simple additions or (2) additions followed by dehydration or other changes.

Simple addition reactions. When ketones and aldehydes are dissolved in water they establish equilibrium with their hydrates, *gem*-diols. This equilibrium is not of much chemical interest, but it does furnish an indication of the ease with which additions can occur across various carbonyl groups.

$$\mathrm{R{-}\overset{O}{\overset{\|}{C}}{-}R'} + \mathrm{H_2O} \rightleftharpoons \mathrm{R{-}\overset{OH}{\overset{|}{\underset{OH}{\underset{|}{C}}}}{-}R'}$$

In Table 6–3 are listed the extent of hydration of a few carbonyl compounds in dilute aqueous solution.

Formaldehyde is almost completely hydrated; with the successive addition of methyl groups one arrives at acetone, negligibly

TABLE 6–3 ■

Compound	% Hydrate at Equilibrium
CH_2O	99.99
CH_3CHO	58
CH_3COCH_3	0
C_6H_5CHO	Slight
CCl_3CHO	100
CF_3COCF_3	100

hydrated. This trend is the result of the stabilizing effect of methyls on double bonds; in this case the electron-feeding methyl groups stabilize the C=O dipole.

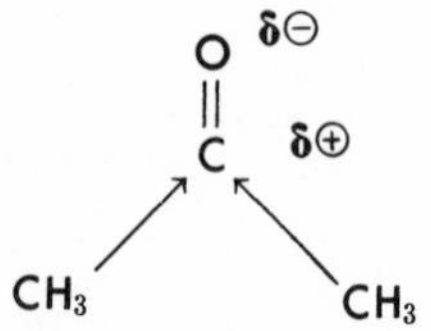

If the carbonyl group is stabilized by conjugation, as in benzaldehyde, there is also very little hydrate formation. Because of this same effect esters, amides, acids, etc., are in general not hydrated to a detectable extent. However, hydrate formation can be promoted by destabilizing the carbonyl form; in chloral (trichloroacetaldehyde) and in hexafluoroacetone the carbon-halogen dipoles interact unfavorably with the carbonyl group.

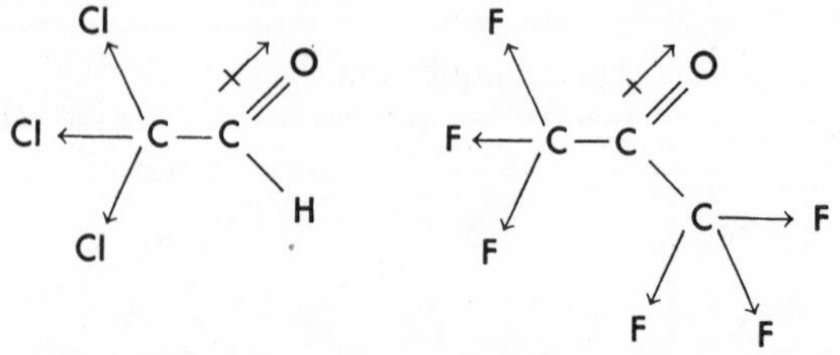

The mechanism by which acetaldehyde reaches its hydration

equilibrium has been studied. The reaction is not instantaneous, but is catalyzed by both general acids and general bases. For the acid-catalyzed process the following mechanism seems reasonable;

$$CH_3{-}CH{=}O + HA \rightleftharpoons CH_3{-}CH{=}O{\cdots}H{-}A \;(+\; H_2\ddot{O}) \xrightarrow{\text{Slow}}$$

$$CH_3{-}CH(OH_2^{\oplus}){-}O{-}H \quad {}^{\ominus}A \xrightarrow{\text{Fast}} CH_3{-}CH(OH){-}OH + HA$$

for the base-catalyzed process it is suggested that the catalyst removes a proton from the attacking water molecule.

$$CH_3{-}CH{=}O + H{-}\ddot{O}{-}H \;\; :B \xrightarrow{\text{Slow}} CH_3{-}CH(OH){-}O^{\ominus} \;\; \overset{\oplus}{B}H \xrightarrow{\text{Fast}} CH_3{-}CH(OH){-}OH \;\; B$$

Although these mechanisms are reasonable and fit the observed catalysis they can hardly be considered as proved. If they are correct, then the principle of microscopic reversibility tells us that the dehydration processes must go back over the same paths. Similar catalysis is found for the hydration of acetone. Although at equilibrium there is a negligible amount of hydrate, the process can be detected by observing O^{18} exchange.

$$CH_3{-}\overset{O^{16}}{\overset{\|}{C}}{-}CH_3 + H_2O^{18} \rightleftharpoons CH_3{-}C(OH)(O^{18}H){-}CH_3 \rightleftharpoons CH_3{-}\overset{O^{18}}{\overset{\|}{C}}{-}CH_3$$

With alcohols, addition to an aldehyde results in hemiacetal formation. The equilibrium extent of this process varies in a manner similar to that for hydration, but it becomes quite important when the alcohol and the aldehyde are in the same molecule, as in glucose and other sugars.

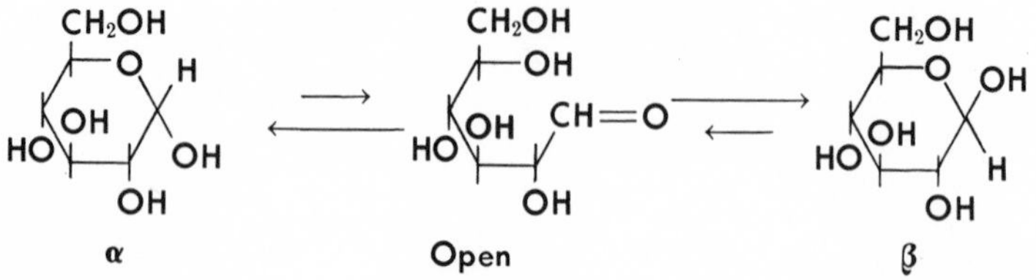

α ⇌ Open ⇌ β

Glucose

The equilibrium lies almost entirely to the two closed forms, called, respectively, α and β,D-glucose. The reason that the equilibrium in hemiacetal formation is so much more favorable here is a matter of entropy. In ordinary hemiacetal formation two molecules, the alcohol and the aldehyde, must be tied down, and there is a considerable loss of freedom of motion; in cyclic hemiacetal formation only one molecule is involved and much less loss of freedom occurs since only some rotational freedom becomes restricted. The rate of ring opening and closing can be studied by following the optical rotation of the solution, since the interconversion of α and β forms involves opening and then closing in the other stereochemical sense. This interconversion, *mutarotation,* is catalyzed by both acids and bases in mechanisms apparently very similar to those written for hydration.

Mercaptans add readily to carbonyl groups, and the simple hemithioacetals and hemithioketals are converted further to thioacetals and thioketals, as discussed in the following section. Furthermore, bisulfite ion and cyanide ion can add reversibly to many carbonyl groups.

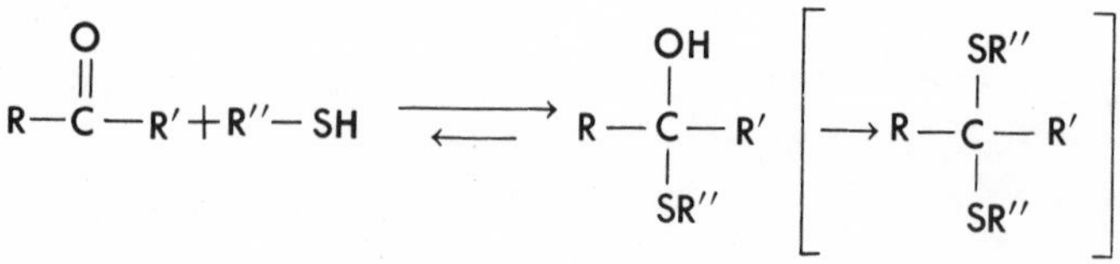

$$R-\overset{O}{\overset{\|}{C}}-R' + R''-SH \rightleftharpoons R-\underset{SR''}{\underset{|}{\overset{OH}{\overset{|}{C}}}}-R' \left[\longrightarrow R-\underset{SR''}{\underset{|}{\overset{SR''}{\overset{|}{C}}}}-R' \right]$$

$$CH_3CHO + HSO_3^{\ominus} \rightleftharpoons CH_3-CH(OH)-SO_3^{\ominus}$$

$$C_6H_5CHO + HCN \rightleftharpoons C_6H_5-CH(OH)-C\equiv N$$

The equilibrium depends on the structure of the carbonyl compound as it did for hydrate formation, adduct formation being less favorable when the carbonyl carries electron-feeding groups.

In the *aldol condensation* an enolate ion adds to a carbonyl group. Thus treatment of acetaldehyde with mild base produces β-hydroxybutyraldehyde, which has the trivial name "aldol." The mechanism currently favored is clearly related to that of the other addition reactions discussed.

$$CH_3CHO + OH^{\ominus} \longrightarrow CH_2{=}CH{-}O^{\ominus} \quad + \quad CH_3-CH{=}O \xrightarrow{\text{Slow}} CH_3-CH(O^{\ominus})-CH_2-CHO \xrightarrow{\text{Fast}} CH_3-CH(OH)-CH_2-CHO$$

When acetaldehyde is present at moderate concentrations the addition step is rate-determining, as the kinetic expression shows.

$$\text{Rate} = k\,(OH^-)\,(CH_3CHO)^2$$

However, when the acetaldehyde concentration is increased considerably every enolate ion gets trapped, as in acetone bromination, and the rate expression becomes first order in acetaldehyde.

$$\text{Rate} = k\,(OH^-)\,(CH_3CHO)$$

In the acid-catalyzed aldol condensation an enol, not an enolate ion, is the species which adds to carbonyl.

A process which formally resembles the aldol condensation occurs when benzaldehyde is treated with alkaline cyanide solution. The product is benzoin, and the *benzoin condensation* involves a specific role for the cyanide ion.

$$2\ C_6H_5CHO \xrightarrow{CN^-} C_6H_5\overset{\overset{O}{\|}}{C}-\underset{\underset{H}{|}}{\overset{\overset{OH}{|}}{C}}-C_6H_5$$

Benzaldehyde cannot of course form an enolate ion, and the hydrogen of the aldehyde group is not acidic. However, when mandelonitrile is formed by addition of HCN across the carbonyl group then hydrogen becomes acidified by the cyano group, and an aldol-like condensation can occur. The full mechanism of this process was presented in Chapter 2.

$$C_6H_5-\underset{\underset{\underset{N}{|||}}{C}}{\overset{\overset{OH}{|}}{C}}-H \xrightarrow{\text{Base}} C_6H_5-\overset{\overset{OH}{|}}{C}^{\ominus}-C\equiv N \longleftrightarrow C_6H_5-\overset{\overset{OH}{|}}{C}=C=N^{\ominus}$$

Complex addition reactions. Under this heading we shall consider reactions in which the addition is followed by dehydration or substitution. For instance, when aldehydes are treated with acidic alcohol, acetals may be formed. The process involves hemiacetal formation followed by a substitution involving a carbonium ion; this is a reversible process whose equilibrium can

$$CH_3\overset{\overset{O}{\|}}{C}H + CH_3OH \rightleftharpoons CH_3-\underset{\underset{OCH_3}{|}}{\overset{\overset{OH}{|}}{C}}-H \overset{H_3O^{\oplus}}{\rightleftharpoons} CH_3-\underset{\underset{OCH_3}{|}}{\overset{\overset{^{\oplus}OH_2}{|}}{C}}-H \overset{-H_2O}{\rightleftharpoons}$$

$$CH_3-\overset{\oplus}{C}(OCH_3)-H \xrightleftharpoons{CH_3OH} CH_3-C(\overset{\oplus}{O}H-CH_3)(OCH_3)-H \rightleftharpoons CH_3-C(OCH_3)(OCH_3)-H$$

be displaced by removal of the water formed. The intermediate carbonium ion is strongly stabilized by conjugation with the unshared electron pairs on oxygen. Thioacetal formation involves a similar mechanism.

The aldol condensation can also be complex. For instance, acid-catalyzed condensation of benzaldehyde with acetophenone leads chiefly to the unsaturated ketone; both the addition and the dehydration involve enolization.

$$C_6H_5\overset{O}{\overset{\|}{C}}CH_3 + H^{\oplus} \rightleftharpoons C_6H_5\overset{\oplus OH}{\overset{\|}{C}}-CH_3 \rightleftharpoons C_6H_5-\overset{OH}{\overset{|}{C}}=CH_2$$

$$C_6H_5CH=O + H^{\oplus} \rightleftharpoons C_6H_5CH=\overset{\oplus}{O}H$$

$$C_6H_5\overset{OH}{\overset{|}{C}H}-CH=\overset{OH}{\overset{|}{C}}-C_6H_5 \rightleftharpoons C_6H_5\underset{OH}{\underset{|}{C}H}-CH_2-\overset{\oplus OH}{\overset{\|}{C}}-C_6H_5 \quad (1)$$

$$\rightleftharpoons$$

$$C_6H_5\overset{\oplus OH_2}{\overset{|}{C}H}-CH=\overset{\cdot OH}{\overset{|}{C}}-C_6H_5 \xrightarrow{2} C_6H_5CH=CH-\overset{\oplus OH}{\overset{\|}{C}}-C_6H_5$$

$$\rightleftharpoons C_6H_5CH=CH-\overset{O}{\overset{\|}{C}}-C_6H_5$$

The reaction is written out in some detail, so that it can be seen that the acid plays several roles. First of all, it catalyzes enolization; it cannot change the equilibrium amount of enol present, of course, since the ketone and the enol have the same number of protons. Second, the acid helps make the benzaldehyde more reactive to addition, and it also makes the hydroxyl a better leaving group by protonating it. Under some conditions the addition step (numbered 1 in the mechanism) is rate-determining, with dehydration being rapid, while in other cases the early steps are all rapidly reversible with dehydration (numbered 2) being rate-controlling. However, the kinetics does not change, since the same rate expression is expected for each situation.

$$\text{Rate} = k(C_6H_5COCH_3)(C_6H_5CHO)(H^+)$$

It will be seen that this rate expression contains the same atoms as are found in the transition state for either step 1 or step 2. This is another example of the problem of kinetic equivalence which was discussed in Chapter 2.

Some important complex addition reactions involve the formation of Schiff bases from the reaction of ketones and aldehydes with amines.

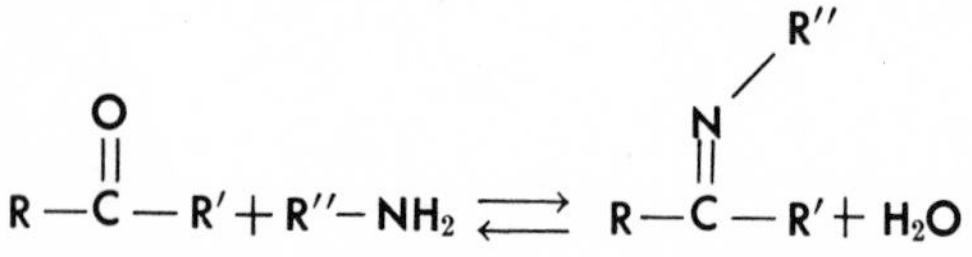

$$R-\overset{\overset{\displaystyle O}{\|}}{C}-R' + R''-NH_2 \rightleftharpoons R-\overset{\overset{\displaystyle N-R''}{\|}}{C}-R' + H_2O$$

With ordinary amines in aqueous solution this equilibrium favors the starting materials, but when the amine component is hydroxylamine or a hydrazine derivative then the imine is stabilized by extra conjugation. With such compounds the reaction goes essentially

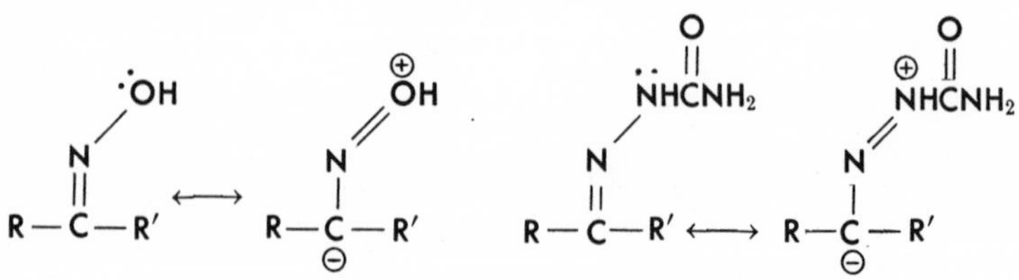

$$R-\overset{\overset{\displaystyle N-\ddot{O}H}{\|}}{C}-R' \longleftrightarrow R-\overset{\overset{\displaystyle N=\overset{\oplus}{O}H}{|}}{\underset{\ominus}{C}}-R'$$

Oxime

$$R-\overset{\overset{\displaystyle N-\ddot{N}HCONH_2}{\|}}{C}-R' \longleftrightarrow R-\overset{\overset{\displaystyle N=\overset{\oplus}{N}HCONH_2}{|}}{\underset{\ominus}{C}}-R'$$

Semicarbazone

to completion and carbonyl derivatives—oximes, hydrazones, semicarbazones—may be isolated.

Oxime formation is a two-step process, addition to the carbonyl followed by dehydration, and in discussing the rate of oxime formation we must know which step is rate-controlling. At a pH above 4 or 5 apparently addition is rapid and dehydration is slow; thus when hydroxylamine is added to furfural the characteristic ultraviolet spectrum of furfural rapidly changes to that of the adduct, and this only slowly goes on by dehydration to form the oxime.

$$\text{C}_4\text{H}_3\text{O-CHO} + H_2NOH \underset{}{\overset{\text{Fast}}{\rightleftharpoons}} \text{C}_4\text{H}_3\text{O-CH(OH)NHOH} \xrightarrow{\text{Slow}} \text{C}_4\text{H}_3\text{O-CH{=}NOH}$$

The initial reactions are just nucleophilic additions of free hydroxylamine, while the dehydrations are acid-catalyzed.

$$CH_3-\overset{O}{\overset{\|}{C}}-CH_3 + :NH_2-OH \rightleftharpoons CH_3-\overset{O^{\ominus}}{\overset{|}{C}}(\oplus NH_2OH)-CH_3 \rightleftharpoons CH_3-\overset{OH}{\overset{|}{C}}(NH-OH)-CH_3 \overset{H^+}{\rightleftharpoons}$$

$$CH_3-\overset{\oplus OH_2}{\overset{|}{C}}(:NH-OH)-CH_3 \xrightarrow{\text{Slow}} CH_3-C(=\oplus NH-OH)-CH_3 \rightleftharpoons CH_3-C(=N-OH)-CH_3$$

Accordingly, it would be expected that oxime formation should be faster in strong acid. Actually the rate does increase on going from pH 7 down to pH 5, but then it drops off again as the pH is further decreased (Figure 6–1).

What happens is as follows: As the pH is lowered the dehydration step gets faster and faster, but the first addition step slows down. The addition requires free hydroxylamine, and in strong acid the hydroxylamine is mostly protonated so that the concentration of *free* hydroxylamine becomes very small. Thus below pH 5 the addition step becomes rate-determining. The addition is of

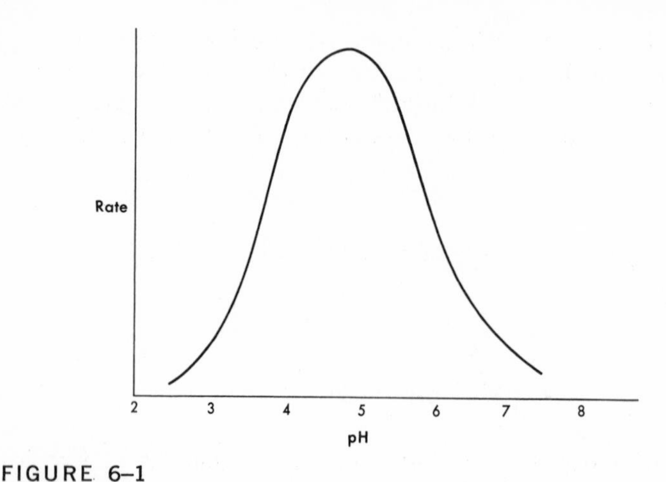

FIGURE 6–1

The rate of oxime formation from acetone as a function of pH.

course slower still in stronger acid because of the low concentration of free hydroxylamine, so the over-all rate falls off at low pH. For this reason oximes and other carbonyl derivatives are often made with an acetic acid catalyst, which produces more or less the optimum pH.

6–3 Reactions of Carboxylic Acids and Their Derivatives

In this section we shall deal first with interconversions among acids, esters, amides, acid chlorides, etc., and then we shall take up more specialized topics such as condensation and decarboxylation. Although perhaps the esterification of a carboxylic acid should be considered first, the principle of microscopic reversibility tells us that the mechanism of esterification is just the reverse of that for ester hydrolysis under the same conditions. Accordingly we shall deal with ester hydrolysis, since this is the process to which most study has been given.

The hydrolysis of esters. When an ester is treated with aqueous acid it hydrolyzes. The reverse process is Fisher esterification, and the equilibrium may be displaced in either direction depending on whether water is present in excess (for hydrolysis) or whether it is removed from the reaction in some way (as is often done in esterifications).

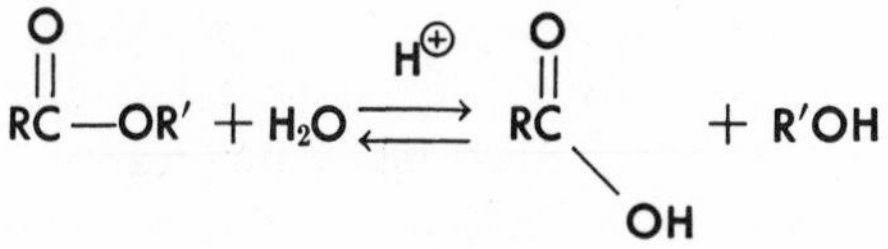

$$RC(=O)—OR' + H_2O \overset{H^{\oplus}}{\rightleftharpoons} RC(=O)OH + R'OH$$

Esters may also be hydrolyzed with base. This process, called saponification ("soap-making"), is generally not reversible since the carboxylic acid forms a salt which displaces the equilibrium.

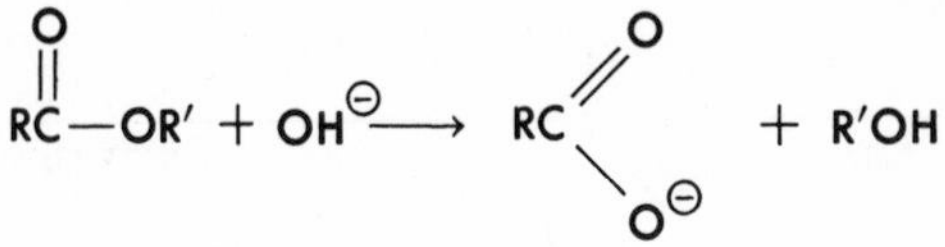

$$RC(=O)—OR' + OH^{\ominus} \longrightarrow RC(=O)O^{\ominus} + R'OH$$

In principle the cleavage of an ester could occur in two different places: between the acyl group and the oxygen, or between the alkyl group and the oxygen. Acyl-oxygen fission is most common, although we shall see some cases in which alkyl-oxygen fission occurs.

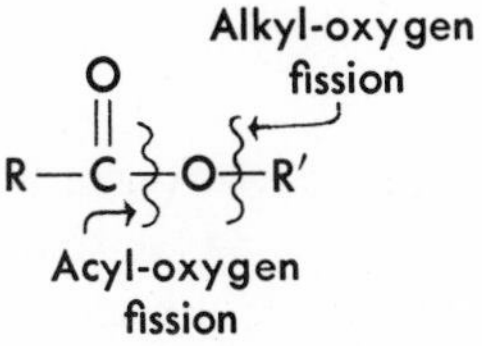

$$R—C(=O) \wr O \wr R'$$

Alkyl-oxygen fission

Acyl-oxygen fission

The most convincing evidence for acyl-oxygen fission in ordinary ester hydrolysis comes from studies with O^{18}. Thus hydrolysis of methyl succinate with H_2O^{18} yields ordinary methanol, the heavy oxygen being found in the succinic acid.

For simplicity the heavy oxygen is written in the OH group of the product carboxyl, although of course rapid proton exchange makes

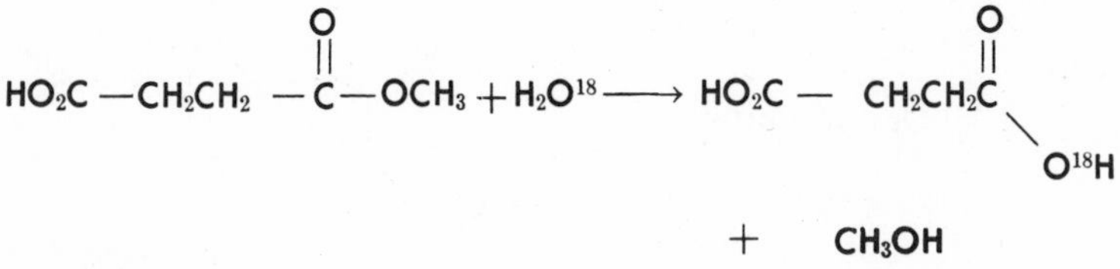

$$HO_2C—CH_2CH_2—C(=O)—OCH_3 + H_2O^{18} \longrightarrow HO_2C—CH_2CH_2C(=O)O^{18}H + CH_3OH$$

the two carboxyl oxygens experimentally equivalent.

Similar experiments have been done for a number of other simple esters. One of the most interesting was the early demonstration that Fisher esterification of benzoic acid with labeled methanol resulted in the formation of labeled methyl benzoate; by the principle of microscopic reversibility this demonstrates that the reverse reaction, hydrolysis, also involves acyl-oxygen fission.

$$C_6H_5C(=O)OH + CH_3O^{18}H \xrightarrow{H^+} C_6H_5C(=O)O^{18}CH_3 + H_2O$$

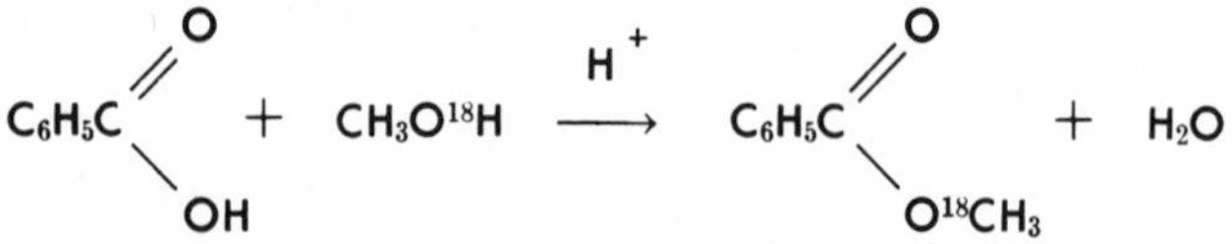

Saponification of ordinary esters also involves acyl-oxygen fission. For instance, saponification of labeled ethyl propionate affords labeled ethanol.

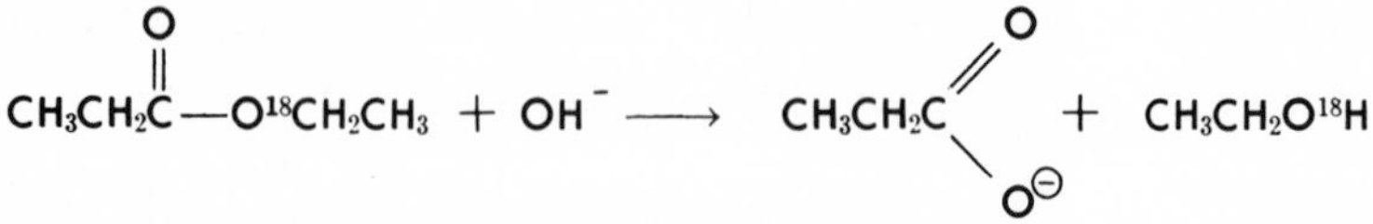

$$CH_3CH_2\overset{O}{\overset{\|}{C}}{-}O^{18}CH_2CH_3 + OH^- \longrightarrow CH_3CH_2C(=O)O^{\ominus} + CH_3CH_2O^{18}H$$

We have emphasized that acyl-oxygen fission is involved in the hydrolysis of "ordinary" esters, so we should mention those cases in which alkyl-oxygen fission occurs instead. Since alkyl-oxygen

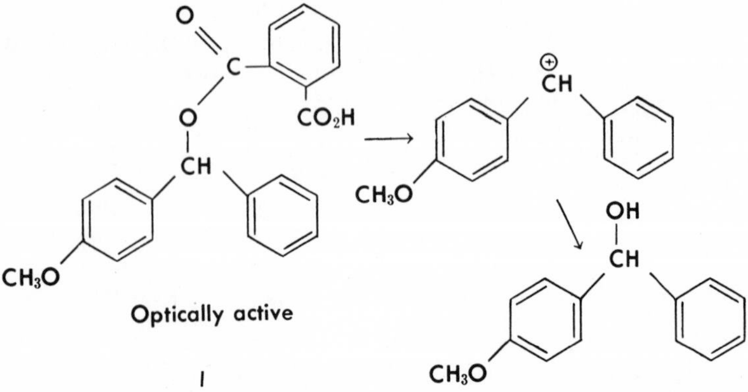

CH₃O–C₆H₄–CH(C₆H₅)–O–C(=O)–C₆H₄–CO₂H → CH₃O–C₆H₄–$\overset{\oplus}{C}H$–C₆H₅ → CH₃O–C₆H₄–CH(OH)–C₆H₅

Optically active

I

Racemic

fission in an ester is simply nucleophilic substitution on the alkyl group, the structural features which make this process especially easy will be factors of the sort already discussed in Chapter 3. For instance, optically active *p*-methoxybenzhydryl phthalate (I) is hydrolyzed to optically inactive benzhydrol even in 10*N* NaOH. The racemization shows that S_N1 substitution is occurring on the alkyl group, and it competes with the ordinary saponification mechanism even in this very strong alkali.

The S_N1 process is highly favorable because an excellent carbonium ion is formed, but a number of other esters undergo this type of hydrolysis in neutral solution. As an example, α-phenethyl phthalate hydrolyzes by an S_N1 mechanism in water, although with base the ordinary saponification mechanism takes over.

$$C_6H_5\overset{*}{C}H(CH_3)-OC(=O)-C_6H_4-CO_2H \xrightarrow{H_2O} \left[C_6H_5\overset{\oplus}{C}H-CH_3 \longrightarrow \right] \longrightarrow C_6H_5CH(CH_3)-OH \text{ (Racemic)}$$

$$C_6H_5\overset{*}{C}H(CH_3)-OC(=O)-C_6H_4-CO_2H \xrightarrow[\text{Saponification}]{OH^-} C_6H_5\overset{*}{C}H(CH_3)-OH \text{ (with retention of optical activity)}$$

Since S_N1 processes are often facilitated by protonation of the leaving group, it might be expected that alkyl-oxygen fission could also occur in some acid-catalyzed ester hydrolyses. This is so when a reasonably stable carbonium ion may be formed, *t*-butyl acetate hydrolyzing by an S_N1 process.

$$(CH_3)_3C-OC(=O)CH_3 + H_2O^{18} \xrightarrow{H^{\oplus}} (CH_3)_3C-O^{18}H + HOC(=O)CH_3$$

In general, only good S_N1 reactions can compete with the ordinary acyl-oxygen fission. S_N2 reactions are too slow, so alkyl-oxygen fission is not seen with methyl esters, for instance. An S_N2 reaction is however involved in the formation of dimethyl ether by the reaction of methoxide ion with methyl benzoate.

$$C_6H_5C(=O)-O-CH_3 + {}^{\ominus}OCH_3 \longrightarrow C_6H_5C(=O)O^{\ominus} + CH_3OCH_3$$

In this case the much faster acyl-oxygen fission reaction is not detected since it simply regenerates starting material.

Returning to the ordinary acyl-oxygen fission reaction of esters, one could imagine two possible mechanisms. In one there is direct displacement of the alkoxide group by hydroxide ion (in saponification), while in the other there is addition of the hydroxide to carbonyl to form a tetrahedral intermediate which then loses alkoxide. Labeling experiments with O^{18} have been used to show that a mechanism involving a tetrahedral intermediate is the correct one.

$$C_6H_5C(=O^{18})-OCH_2CH_3 + OH^- \longrightarrow \underset{\text{I}}{C_6H_5-C(O^{18\ominus})(OH)-OCH_2CH_3} \longrightarrow$$

$$C_6H_5C(=O)OH + {}^{\ominus}OCH_2CH_3 \longrightarrow C_6H_5C(=O)O^{\ominus} + CH_3CH_2OH$$

The experiment was performed by running the hydrolysis only part way, and then examining the recovered ester. If the tetrahedral intermediate (I) is formed reversibly, and if protons rapidly come on and off hydroxyl groups (as they do), then one might expect O^{18} to be lost from the ester. This was the result actually observed.

$$\underset{\text{I}}{C_6H_5-C(O^{18\ominus})(OH)-OEt} \rightleftarrows C_6H_5-C(O^{18}H)(O^{\ominus})-OEt \longrightarrow C_6H_5C(=O)OEt + {}^{\ominus}O^{18}H$$

The amount of O^{18} exchange observed depends on the relative rate at which the tetrahedral intermediate (I) loses hydroxide ion versus ethoxide ion. In this particular case hydrolysis is 11 times as fast as oxygen exchange in the recovered ester, so that most of the tetrahedral intermediate molecules lose ethoxide ion.

The same O^{18} exchange was observed for the acid hydrolysis of ethyl benzoate, showing that it also involves the formation of a tetrahedral intermediate.

$$C_6H_5C(=O)-OEt + H_2O \underset{}{\overset{H^+}{\rightleftharpoons}} C_6H_5C(=\overset{\oplus}{O}H)-OEt \rightleftharpoons$$

$$C_6H_5-C(OH)(\overset{\oplus}{O}H_2)-OEt \rightleftharpoons C_6H_5-C(OH)(OH)-OEt \rightleftharpoons$$

$$C_6H_5-C(OH)(OH)-\overset{\oplus}{O}(H)Et \rightleftharpoons C_6H_5C(=\overset{\oplus}{O}H)(OH) + EtOH \rightleftharpoons$$

$$C_6H_5C(=O)(OH) + EtOH$$

The mechanism is written out in great detail, showing each individual protonation step, to illustrate the fact that it is completely symmetrical. Run backwards it is an acid-catalyzed esterification, and it is apparent that the hydrolysis and the esterification have exactly the same mechanisms. The oxygen exchange occurs in the reversible formation of the tetrahedral intermediate.

$$C_6H_5C(=O^{18})OEt + H_2O \underset{\text{Steps}}{\overset{\text{Several}}{\rightleftharpoons}} C_6H_5-C(O^{18}H)(OH)-OEt \rightleftharpoons C_6H_5C(=O)OEt + H_2O^{18}$$

The hydrolysis of other carboxyl derivatives. Amides may be hydrolyzed by either acid or base, although acid hydrolysis is usually more rapid. In the basic hydrolysis a tetrahedral intermediate has been demonstrated by O^{18} exchange, and the exchange is actually faster than hydrolysis, NH_2^- being such a poor leaving group.

$$C_6H_5C(=O^{18})-NH_2 \underset{}{\overset{OH^-}{\rightleftharpoons}} C_6H_5C(O^{\ominus})(OH)-NH_2 \rightleftharpoons C_6H_5C(=O)NH_2 + {}^{\ominus}O^{18}H$$

$$C_6H_5C(O^{\ominus})(OH)-NH_2 \downarrow \quad C_6H_5C(=O)OH + {}^{\ominus}NH_2 \longrightarrow C_6H_5C(=O)O^{\ominus} \quad NH_3$$

The acid hydrolysis probably also proceeds through a tetrahedral intermediate, but in this case no O^{18} exchange is found in the recovered amide since a much better leaving group, neutral ammonia, is involved in the forward reaction.

$$C_6H_5C(=O)-NH_2 \longrightarrow C_6H_5C(={}^{\oplus}OH)NH_2 \longrightarrow C_6H_5-C(OH)(^{\oplus}OH_2)-NH_2 \longrightarrow$$

$$C_6H_5-C(OH)_2-\overset{\oplus}{N}H_3 \longrightarrow C_6H_5C(={}^{\oplus}OH)OH + NH_3 \longrightarrow C_6H_5C(=O)OH + NH_4^{\oplus}$$

Some oxygen exchange has even been found in the neutral hydrolysis of benzoyl chloride and of benzoic anhydride, showing that these compounds also form a tetrahedral intermediate. However, most of the intermediate goes on to product, since chloride ion and benzoate ion are such good leaving groups.

$$C_6H_5C(=O)-Cl + H_2O \rightleftharpoons C_6H_5C(OH)_2-Cl \longrightarrow$$

$$C_6H_5C(={}^{\oplus}OH)OH + Cl^{\ominus} \longrightarrow C_6H_5C(=O)OH$$

$$C_6H_5C(=O)—OC(=O)C_6H_5 \rightleftharpoons C_6H_5—C(OH)(OH)—OC(=O)C_6H_5 \longrightarrow$$

$$2\ C_6H_5C(=O)OH$$

It should be noted in passing that we can explain why the postulated tetrahedral intermediates in these cases give only a small amount of exchange, but we cannot rule out the possibility that a tetrahedral intermediate is not formed at all in the cases in which exchange is not observed. Even the *observed* exchange reactions could be side processes unrelated to the hydrolysis mechanism. The carbonyl addition mechanism seems an attractive idea, but it may not be correct in all cases.

The Claisen condensation. When ethyl acetate is treated with sodium ethoxide it is transformed into acetoacetic ester. This condensation reaction between two molecules of an ester is called the Claisen condensation.

$$2CH_3C(=O)OEt \xrightarrow{NaOEt} CH_3C(=O)CH_2C(=O)OEt \xrightarrow{OEt^-} CH_3C(O^{\ominus})=CH—C(=O)OEt$$

Since a β-ketoester is quite acidic, the product actually formed is the enolate salt. The reaction involves attack by the enolate ion of one ethyl acetate molecule on the carbonyl of a second; the resulting tetrahedral species then loses ethoxide ion.

$$CH_3—C(=O)OEt + OEt^{\ominus} \rightleftharpoons CH_2=C(O^{\ominus})OEt$$

$$CH_3-\overset{O}{\overset{\|}{C}}-OEt + CH_2{=}\underset{O^{\ominus}}{\underset{|}{C}}-OEt \rightleftarrows CH_3-\underset{OEt}{\underset{|}{\overset{O^{\ominus}}{\overset{|}{C}}}}-CH_2\overset{O}{\overset{\|}{C}}-OEt \rightleftarrows$$

$$CH_3-\overset{O}{\overset{\|}{C}}-CH_2\overset{O}{\overset{\|}{C}}OEt$$

As shown, the process is reversible and the reaction is successful only because the product is converted to a stable enolate, so that the equilibrium is displaced.

Decarboxylation of acids. The loss of a carboxyl group is similar to the loss of a proton.

$$R-C(=O)-O-H \longrightarrow R^{\ominus} \quad O{=}C{=}O \quad H^{\oplus}$$

$$R-H \longrightarrow R^{\ominus} \quad H^{\oplus}$$

In both processes the electron pair which originally bound the carboxyl or proton to R must stay behind; decarboxylation will occur easily only when the resulting species is stabilized.

An obvious example is found in the decarboxylation of β-ketoacids such as acetoacetic acid. Here kinetics shows that both the free acid and the carboxylate ion can decarboxylate.

$$\text{Rate} = k_1\ (CH_3COCH_2COOH) + k_2\ (CH_3COCH_2COO^-)$$

From the carboxylate ion an enolate results, while the free acid affords the enol of acetone.

$$CH_3-\overset{O}{\overset{\|}{C}}-CH_2-\overset{O}{\overset{\|}{C}}-O^{\ominus} \longrightarrow CH_3-\overset{O^{\ominus}}{\overset{|}{C}}{=}CH_2 + CO_2 \longrightarrow CH_3\overset{O}{\overset{\|}{C}}CH_3$$

$$CH_3-\overset{O}{\overset{\|}{C}}-CH_2-\overset{O-H}{C}=O \longrightarrow CH_3-\overset{OH}{C}=CH_2 + CO_2 \longrightarrow CH_3\overset{O}{\overset{\|}{C}}CH_3$$

These are processes in which the carboxyl group is lost first, and then the proton adds in its place. In principle there are two other possibilities as well; a proton could simultaneously replace a carboxyl group, or it could come in before the carboxyl is lost. There is no known example of the first reaction, which would be a bimolecular electrophilic substitution, although in organometallic chemistry bimolecular electrophilic reactions are common.

$$R-C(=O)-O-H + H^{\oplus} \not\longrightarrow R-H + O=C=O + H^{\oplus}$$

On the other hand, a number of examples are known in which

$$\text{anthracene-9-}CO_2H \xrightarrow{H^{\oplus}} \text{(9-H, 9-COOH, 10-}\oplus\text{ cation)} \longrightarrow \text{anthracene} + CO_2 + H^{\oplus}$$

protonation precedes decarboxylation. The decarboxylation of anthracene-9-carboxylic acid is such a case; it will be seen that this is really electrophilic substitution on an aromatic ring.

In Chapter 7 we shall also come across an example of free radical decarboxylation; in Special Topic 6 we shall mention an example of oxidative decarboxylation.

General References

C. Gutsche, *The Chemistry of Carbonyl Compounds* (Prentice-Hall, Englewood Cliffs, New Jersey, 1967). Emphasizes synthetic methods rather than evidence on mechanisms.

J. Hine, *Physical Organic Chemistry* (2nd ed., McGraw-Hill Book Co., New York, 1962), Chapters 10, 11, 12, and 13. The best general reference for this material.

M. Newman, "Additions to Unsaturated Functions," in M. Newman, ed., *Steric Effects in Organic Chemistry* (John Wiley & Sons, New York, 1956). A good discussion of mechanisms which emphasizes steric hindrance effects on reaction rates of ketones, esters, etc.

M. Bender, "Mechanisms of Catalysis of Nucleophilic Reactions of Carboxylic Acid Derivatives," *Chemical Reviews,* **60,** 53 (1960). There is a good review of the mechanisms of hydrolysis of esters in this article; most of it is concerned with catalysis, including catalysis by enzymes and models for enzymes.

R. Bell, *The Proton in Chemistry* (Cornell University Press, Ithaca, New York, 1959). The later chapters discuss enolization, and provide useful rate data.

D. Cram, *Fundamentals of Carbanion Chemistry* (Academic Press, New York, 1965). Some of the carbanions discussed are enolate ions.

6

Special Topic

▪ OXIDATION-REDUCTION REACTIONS[1]

OXIDATIONS AND REDUCTIONS occur quite generally throughout organic chemistry, but it seems appropriate to examine them formally at this point since so many of the interesting cases involve carbonyl compounds.

It is actually rather difficult to define oxidation in a satisfactory way. It is often considered to be electron removal, but this concept is not useful when atoms as well as electrons are changing place. The following example will illustrate the dilemma.

$$CH_2{=}CH_2 + Br_2 \longrightarrow \underset{Br\quad\; Br}{CH_2{-}CH_2} \xrightarrow{OH^-} \overset{OH\quad OH}{CH_2{-}CH_2} + 2Br^-$$

1 (a). The best general review available in this field is R. Stewart, *Oxidation Mechanisms* (W. A. Benjamin, New York, 1964). (b). Review articles at an advanced level are found in *Oxidation in Organic Chemistry,* K. Wiberg, ed. (Academic Press, New York, 1965).

The over-all process is clearly an oxidation-reduction, for the original bromine molecule has been reduced to two bromide ions. The second reaction written is just nucleophilic displacement, so it is presumably in the first step that the oxidation-reduction occurs. This is sensible if we adopt the arbitrary idea that in a carbon-bromine bond the electrons already "belong" to the bromine, for then the first reaction is indeed an electron transfer. However, this definition does depend on quite an arbitrary decision, and it is apparent that this particular oxidation reaction is a perfectly normal ionic addition to a double bond.

We have already discussed a number of oxidation mechanisms in Special Topic 4. Thus, ozonization of a double bond, epoxidation of an olefin with a peracid, and addition of osmium tetroxide to an olefin are oxidation reactions as well as being cycloaddition reactions. A similar cyclic mechanism has been demonstrated for the hydroxylation of olefins by alkaline permanganate; O^{18} labeling experiments show that both oxygens of the resulting glycol come from the permanganate ion.[2]

$$\text{cyclopentene} + \left[\mathrm{MnO_4}^{18}\right]^{-} \longrightarrow \text{(cyclic ester: C–O, C–O} \overset{\ominus}{\mathrm{MnO_2}}\text{)} \longrightarrow \text{cyclopentane-1,2-diol } (\mathrm{O^{18}H}, \mathrm{O^{18}H}) + \mathrm{Mn\,V}$$

The intermediate cyclic manganate ester cannot be isolated. It rapidly hydrolyzes to the glycol in strong base, although at lower pH it may be further oxidized by more permanganate ion and the yield of glycol is lowered.[3]

It must not be thought that organic oxidations and reductions never involve simple electron transfer. Oxidizing agents such as ferricyanide ion can remove one electron from the hydroquinone dianion to form semiquinone, while alkali metals can add one electron to a variety of unsaturated systems.

2. K. Wiberg and K. Saegebarth, "The Mechanisms of Permanganate Oxidation. IV. Hydroxylation of Olefins and Related Reactions," *Journal of the American Chemical Society*, **79,** 2822 (1957).
3. Ref. 1, p. 62.

$$^{\ominus}O-C_6H_4-O^{\ominus} + Fe^{III}(CN)_6^{\equiv} \longrightarrow {}^{\odot}O-C_6H_4-O^{\ominus} + Fe^{II}(CN)_6^{\equiv}$$

$$C_6H_5\overset{O}{\overset{\|}{C}}C_6H_5 + K\cdot \longrightarrow C_6H_5-\underset{\odot}{\overset{O^{\ominus}}{\overset{|}{C}}}-C_6H_5 \quad K^{\oplus}$$

The free radicals in the cases shown are stabilized by conjugation. Furthermore, in Kolbe electrolysis[4] an electron is simply removed from a carboxylate ion, although the resulting radical then undergoes further changes.

$$CH_3C\begin{matrix}\nearrow O \\ \searrow O^{\ominus}\end{matrix} \xrightarrow{-e} CH_3-C\begin{matrix}\nearrow O \\ \searrow O\cdot\end{matrix} \longrightarrow CH_3\cdot + CO_2$$

$$2\ CH_3\cdot \longrightarrow CH_3-CH_3$$

However, since most work on mechanisms has concerned two general groups of reactions—hydride transfers, and oxidations by inorganic compounds—we shall discuss examples of these in more detail.

Hydride Transfer Reactions

The reduction of a carbonyl group by lithium aluminum hydride

$$R-\overset{O}{\overset{\|}{C}}-R + H-\overset{H}{\overset{|}{\underset{H}{\underset{|}{B}}}}{}^{\ominus}-H \longrightarrow R-\overset{O^{\ominus}}{\overset{|}{\underset{H}{\underset{|}{C}}}}-R + BH_3 \longrightarrow R-\overset{OH}{\overset{|}{\underset{H}{\underset{|}{C}}}}-R$$

4. Ref. 1, p. 128.

or sodium borohydride involves transfer of a hydride ion.[5] Hydride ions can also be donated by organic molecules. Thus, treatment of benzaldehyde with strong base causes an oxidation-reduction, the Cannizzaro reaction.[6]

$$C_6H_5-C(=O)H + OH^- \rightleftharpoons C_6H_5-C(O^\ominus)(OH)-H \longrightarrow$$

$$H-C(=O)-C_6H_5$$

$$C_6H_5-C(=O)OH + C_6H_5-C(H)(O^\ominus)-H \longrightarrow C_6H_5CO_2^\ominus + C_6H_5CH_2OH$$

As required by this mechanism the reaction is second order in benzaldehyde and first order in hydroxide ion; when the reaction is run in D_2O as solvent the benzyl alcohol formed still has only hydrogen on carbon, as is expected. With furfural, the reaction becomes fourth order kinetically, second order in both aldehyde and hydroxide.[6] Here the carbonyl is less reactive, and a stronger hydride donor is required.

$$C_4H_3O-CHO \overset{OH^\ominus}{\rightleftharpoons} C_4H_3O-C(OH)(O^\ominus)-H \overset{OH^\ominus}{\rightleftharpoons} C_4H_3O-C(O^\ominus)(O^\ominus)-H$$

$$C_4H_3O-CH=O \longleftarrow \longleftarrow$$

$$C_4H_3O-CO_2^\ominus + C_4H_3O-CH_2OH$$

5. N. Gaylord, *Reduction with Complex Metal Hydrides* (Interscience Publishers, New York, 1956).
6. J. Hine, *Physical Organic Chemistry* (2nd ed., McGraw-Hill Book Co., New York, 1962), p. 267.

Any alkoxide ion can be a hydride donor if there is an adjacent hydrogen. However, aluminum alkoxides prove to be particularly effective in reducing other carbonyl compounds. The use of a compound such as aluminum isopropoxide to catalyze the reduction of benzaldehyde by isopropanol is an example of the Meerwein-Ponndorf reaction, if it is considered to be a reduction of benzaldehyde, or the Oppenauer reaction if it is considered to be an oxidation of isopropanol. The reaction probably involves a cyclic transition state with transfer of a hydride ion.[7]

$$C_6H_5CHO + CH_3-\overset{\displaystyle OH}{\overset{|}{C}H}-CH_3 \underset{\longleftarrow}{\overset{Al(OiPr)_3}{\longrightarrow}} C_6H_5CH_2OH + CH_3\overset{\displaystyle O}{\overset{\|}{C}}CH_3$$

$$\text{Al}\;\left[\;C_6H_5-\underset{H}{\overset{O}{C}}\;\;\;\; H-\overset{O}{C}(CH_3)_2\;\right] \longrightarrow \text{Al}\;\left[\;C_6H_5-\underset{H}{\overset{O}{C}}-H \;\;\;\; \overset{O}{\overset{\|}{C}}(CH_3)_2\;\right]$$

A similar cyclic transition state is involved in the abnormal Grignard reaction, a reduction which occurs with some hindered ketones.[8]

$$CH_3-CH_2-CH_2MgBr + (CH_3)_2CH-\overset{\displaystyle O}{\overset{\|}{C}}-CH(CH_3)_2 \longrightarrow$$

$$(CH_3)_2CH-\underset{\displaystyle H}{\underset{|}{\overset{\displaystyle OH}{\overset{|}{C}}}}-CH(CH_3)_2 + CH_3-CH=CH_2 + \text{Normal adduct}$$

7. Ref. 1, p. 19.
8. M. Kharasch and O. Reinmuth, *Grignard Reactions of Nonmetallic Substances* (Prentice-Hall, New York, 1954), p. 147.

Br | Mg ... O=C< ... H ... C(H)(CH$_3$)—CH$_2$ $\longrightarrow$ —C(H)—O—Mg—Br + $CH_2{=}CH{-}CH_3$

These have been examples of hydride transfer within a complex; other cases are known in which true intramolecular hydride transfer occurs. One case which has been studied is the rearrangement of phenylglyoxal to mandelic acid.[6]

$$C_6H_5-\overset{O}{\overset{\|}{C}}-CHO + OH^{\ominus} \longrightarrow C_6H_5-\overset{OH}{\overset{|}{C}H}-CO_2^{\ominus}$$

When this reaction is run in D_2O no carbon-bound deuterium is found in the product (the mandelate anion has an enolizable hydrogen, but exchange next to a carboxylate ion is quite slow). Thus the mechanism involves an intramolecular hydride shift, as shown.

$$C_6H_5\overset{O}{\overset{\|}{C}}-CH{=}O + OH^{\ominus} \rightleftharpoons C_6H_5-\overset{O}{\overset{\|}{C}}-\overset{OH}{\overset{|}{\underset{H}{\underset{|}{C}}}}-O^{\ominus} \longrightarrow$$

$$C_6H_5\overset{O^{\ominus}}{\overset{|}{\underset{H}{\underset{|}{C}}}}-C\langle^{OH}_{O} \longrightarrow C_6H_5\overset{OH}{\overset{|}{C}H}-CO_2^{\ominus}$$

Such a reaction is very similar to the benzilic acid rearrangement, in which a phenyl migrates when benzil is treated with base.[9]

In fact, the rearrangement of phenylglyoxal to mandelic acid

9. S. Selman and J. Eastham, "Benzilic Acid and Related Rearrangements," *Quarterly Reviews*, **14**, 221 (1960).

$$C_6H_5\overset{O}{\overset{\|}{C}}-\overset{O}{\overset{\|}{C}}C_6H_5 + OH^{\ominus} \longrightarrow HO-\underset{C_6H_5}{\overset{O^{\ominus}}{\overset{|}{C}}}-\overset{O}{\overset{\|}{C}}-C_6H_5 \longrightarrow$$

Benzil

$$HO\overset{O}{\overset{\|}{C}}-\underset{C_6H_5}{\overset{O^{\ominus}}{\overset{|}{C}}}-C_6H_5 \longrightarrow (C_6H_5)_2\overset{OH}{\overset{|}{C}}H-CO_2^{\ominus}$$

could actually have involved a phenyl shift; this has been excluded by labeling experiments using C^{14}.

$$C_6H_5\overset{O}{\overset{\|}{C}}{}^*-CHO \xrightarrow{OH^{\ominus}} C_6H_5-\overset{OH}{\overset{|*}{C}}H-CO_2H \left[\xrightarrow{\text{Oxidize}} C_6H_5\overset{*}{C}O_2H + CO_2\right]$$

So far we have been examining cases in which hydride ions are transferred, to carbonyl groups, from molecules which are especially good hydride donors. There are also cases known in which an especially good hydride acceptor may pull hydride ion from an otherwise unreactive molecule. The most important examples of this type of process are found in carbonium ion chemistry. A carbonium ion can usually abstract hydride from a hydrocarbon, particularly if this leads to a better carbonium ion.[10]

$$R\oplus + H-R' \rightleftarrows R-H + R'\oplus$$

For instance, tropylium ion may be synthesized by treating cycloheptatriene with triphenylmethyl cation.[11] The aromatic tropylium ion is a much better carbonium ion, so the reaction goes to completion.

10. N. Deno, H. Peterson, and G. Saines, "The Hydride-Transfer Reaction," *Chemical Reviews,* **60,** 7 (1960).
11. H. Dauben, F. Gadecki, K. Harmon, and D. Pearson, "Synthesis of Tropenium (Cycloheptatrienylium) Salts by Hydride Exchange," *Journal of the American Chemical Society,* **79,** 4557 (1957).

$$(C_6H_5)_3C^{\oplus}\ BF_4^{\ominus} + C_7H_8 \longrightarrow (C_6H_5)_3CH + C_7H_7^{\oplus}\ BF_4^{\ominus}$$

Oxidation by Inorganic Compounds

We have already mentioned the oxidation of olefins by ozone, MnO_4^-, OsO_4, etc. One of the most useful oxidizing agents for the conversion of alcohols to ketones is chromic acid, commonly used in acetic acid solution (although another useful combination is pyridine and CrO_3). Westheimer and his research group have studied the oxidation of isopropyl alcohol to acetone by chromic acid solutions. The mechanism involved is apparently as follows.[12]

$$(CH_3)_2CH{-}OH + HCrO_4^- + 2H^+ \rightleftharpoons (CH_3)_2C(H){-}O{-}\overset{\oplus}{Cr}(=O)(OH)_2 \xrightarrow{:B} (CH_3)_2C{=}O + \overset{\oplus}{B}H + O{=}Cr(OH)_2$$

A chromate ester is reversibly formed, and in the rate-determining step there is an elimination with loss of a Cr^{IV} species. Through a series of other rapid reactions this is converted to Cr^{III}, the stable final oxidation state.

Evidence that a proton is being removed in the rate-determining step of the oxidation, as the mechanism indicates, is found in the kinetic isotope effect observed when deuterioisopropanol, $(CH_3)_2CDOH$, is oxidized. The deuterio compound is 6.6 times slower at 25°C than normal isopropanol; such large

12. F. Westheimer, "The Mechanisms of Chromic Acid Oxidations," *Chemical Reviews,* **45,** 419 (1949); cf. also ref. 1, p. 33.

isotope effects always indicate[13] that the bond to hydrogen is being broken in the transition state. Another piece of evidence in favor of this mechanism is the observation[12] that diisopropyl chromate will undergo base-catalyzed elimination with pyridine, a process very similar to the elimination written for the normal reaction.

$$(CH_3)_2CH-O-\underset{O}{\overset{O}{\overset{\|}{\underset{\|}{Cr}}}}-O-CH(CH_3)_2 \xrightarrow{C_5H_5N:} CH_3\overset{O}{\overset{\|}{C}}CH_3 + Cr^{IV}\ \text{Species}$$

Aldehydes may also be oxidized with chromic acid, and the mechanism is apparently quite similar to that already discussed.[14]

$$RCHO + H^{\oplus} + HCrO_4^{\ominus} \rightleftharpoons R-\underset{O-CrO_2-OH}{\overset{OH}{\overset{|}{\underset{|}{C}}}}-H \xrightarrow{:B} R-C\overset{OH}{\underset{O}{\langle}} + Cr^{IV}\text{, etc.}$$

An isotope effect on the reaction rate of 4.3 was observed for the oxidation of C_6H_5CDO. Again this 4-fold slower reaction for the deuterium compounds shows that the proton is being removed in the rate-determining step. Many saturated hydrocarbons can also be oxidized by chromic acid, as well as by permanganate ion.[15] In these cases apparently a hydrogen atom is removed to form a free radical, which undergoes further oxidation by the inorganic species.

When 1,2-diols are treated with periodic acid they are cleaved

13. K. Wiberg, "The Deuterium Isotope Effect," *Chemical Reviews,* **55,** 713 (1955).
14. Ref. 1, p. 48.
15. Ref. 1, p. 50.

oxidatively.[16] Thus ethylene glycol affords formaldehyde and iodic acid, and the mechanism involves formation of a cyclic periodate.

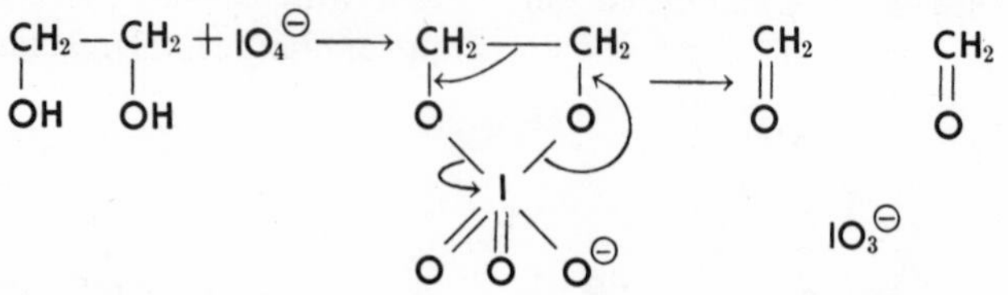

$$\underset{OH}{\underset{|}{CH_2}}-\underset{OH}{\underset{|}{CH_2}} + IO_4^{\ominus} \longrightarrow \text{cyclic periodate ester } (CH_2-CH_2,\ O-I(=O)_2(O^{\ominus})-O) \longrightarrow \underset{O}{\underset{\|}{CH_2}} \quad \underset{O}{\underset{\|}{CH_2}} \quad IO_3^{\ominus}$$

O^{18} labeling experiments show that the carbonyl oxygens of the products come from the original glycol oxygens. One of the pieces of evidence in favor of the cyclic mechanism is the fact that glycols for which such a cyclic ester would be impossible, such as *trans*-9,10-decalindiol, are unaffected by periodate.

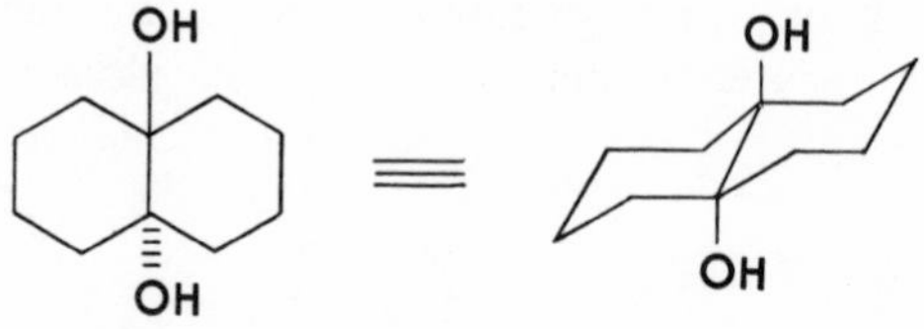

OH OH ≡ OH OH

α-Diketones are also cleaved by periodate, and again a cyclic mechanism seems probable.

Labeling experiments show that the two new oxygens in the product acetic acid molecules in fact come from the periodate, as this mechanism suggests.

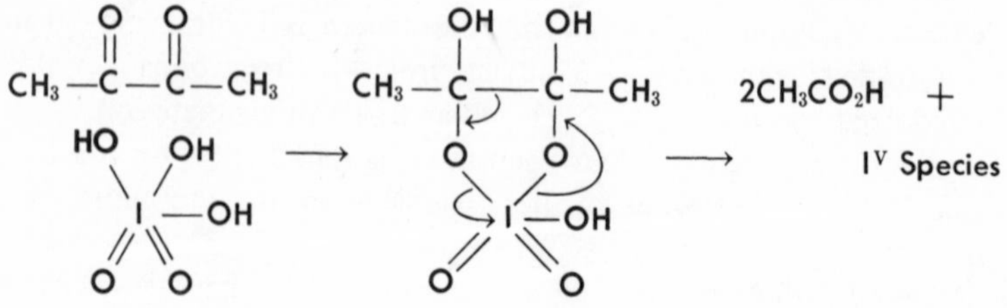

$$CH_3-\overset{O}{\overset{\|}{C}}-\overset{O}{\overset{\|}{C}}-CH_3 \quad (HO)_2I(OH)(=O)_2 \longrightarrow CH_3-\overset{OH}{\overset{|}{C}}-\overset{OH}{\overset{|}{C}}-CH_3 \text{ (cyclic ester with } O-I(OH)(=O)_2-O) \longrightarrow 2CH_3CO_2H + I^{V}\text{ Species}$$

Lead tetraacetate can also cleave glycols, although it is less

16. Ref. 1, Chapter 7.

selective than periodate.[16] There are some data in favor of a cyclic mechanism for this reaction as well, but the cleavage of *trans*-9,10-decalindiol shows that a noncyclic mechanism is also possible.

$$\begin{matrix} -\mathrm{C}-\mathrm{OH} \\ -\mathrm{C}-\mathrm{OH} \end{matrix} + \mathrm{Pb(OAc)_4} \longrightarrow \begin{matrix} -\mathrm{C}-\mathrm{O} \\ -\mathrm{C}-\mathrm{O} \end{matrix} \mathrm{Pb(OAc)_2} \longrightarrow \begin{matrix} -\mathrm{C}{=}\mathrm{O} \\ -\mathrm{C}{=}\mathrm{O} \end{matrix} + \mathrm{Pb(OAc)_2}$$

$$\text{H—O (decalin) O—Pb(OAc)}_3 \longrightarrow \text{cyclodecane-1,6-dione} + \mathrm{Pb(OAc)_2} + \mathrm{AcOH}$$

Thus it is apparent that for many of these inorganic oxidations a standard mechanism is involved: an inorganic ester is formed between the oxidizing agent and the substrate, and then an ordinary elimination reaction or cyclo-decomposition process leads to products. However, when the substrate is a hydrocarbon rather than an alcohol or ketone then free radical chain processes become more important. Such processes are discussed in the next chapter.

7

REACTIONS INVOLVING FREE RADICALS

A FREE RADICAL is any species which has an unpaired electron. Under this definition one would include not only such molecules as NO and NO_2 (which have an odd number of electrons so that one must be unpaired) but also atoms such as I· and Na·. O_2 is also included since it has two unpaired electrons and is thus a biradical. A few radicals, such as the nitrogen oxides above, are stable species; most free radicals are not stable, but they may still play an important role as reactive intermediates. Typically, reactions at high temperatures, including reactions in flames, involve free radical intermediates. Furthermore, many organic reactions at ordinary temperatures go by free radical mechanisms; this is particularly true of reactions in the gas phase or in nonpolar solvents. Before we consider such reactions in detail we will mention briefly the factors which can lead to the stable existence of organic radicals.

7–1 Stable Organic Radicals

Hexaphenylethane is in equilibrium with the triphenylmethyl radical in solution.

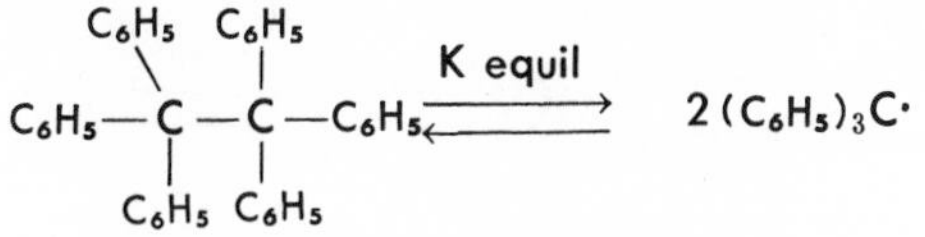

$$(C_6H_5)_3C{-}C(C_6H_5)_3 \overset{K\ \text{equil}}{\rightleftharpoons} 2\,(C_6H_5)_3C\cdot$$

Although this equilibrium is quite unfavorable (a 0.1*M* solution of hexaphenylethane is 2% dissociated at 20°C), the triphenylmethyl radical is easily detected in solutions of hexaphenylethane, and the tris-*p*-nitro derivative exists almost completely dissociated.

A major part of the stabilization of these radicals comes from extra resonance forms which are not available in the dimers.

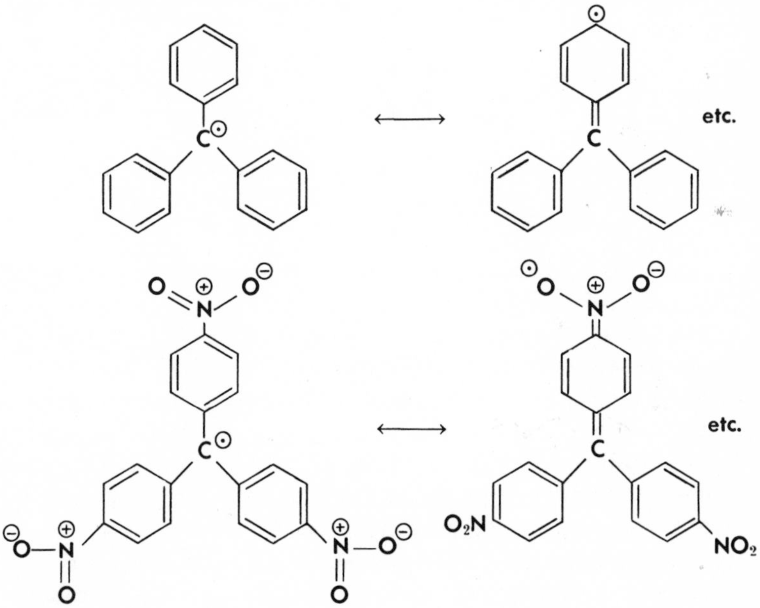

etc.

O_2N NO_2

etc.

The participation of the nitro groups in delocalizing the odd electron explains the greater stability of tris-*p*-nitrophenylmethyl radical. Dissociation is also favored by steric crowding in the dimer, the hexaphenylethane. The three bulky phenyl groups on one carbon are jammed into the three on the other carbon of the ethane.

A number of other stable free radicals are known. Compounds like diphenylpicrylhydrazyl (I) are stable chiefly because of extensive delocalization of the odd electron, while a species such as the anthracene anion radical (II) fails to dimerize both because of conjugative stabilization in the radical and because of electrostatic repulsion of one anion molecule by another.

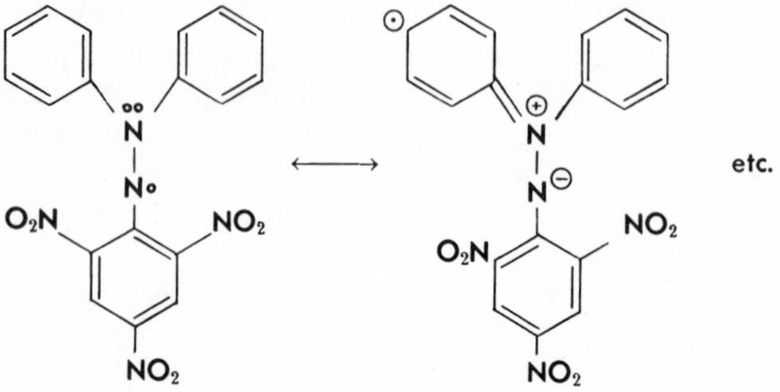

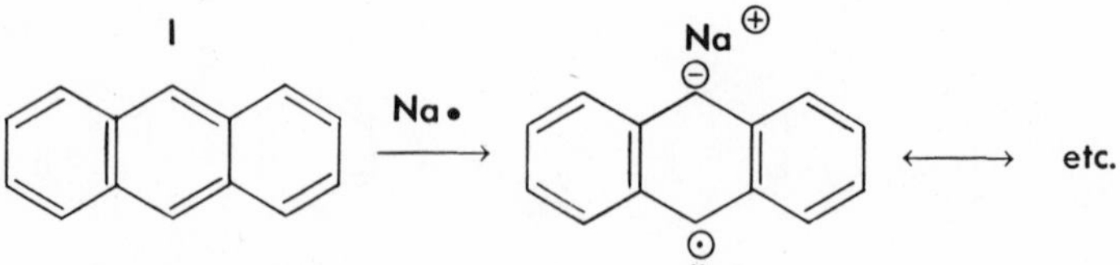

A free radical has a net magnetic moment. All electrons have spin, and a spinning charge generates a magnetic field, but in ordinary molecules the electrons are paired and the magnetic field from those electrons with spin "up" is exactly cancelled by the field from electrons with spin "down." Since a radical has an odd number of electrons there will be one net spin uncompensated, so a free radical generates a magnetic field. Thus free radicals are attracted by other magnetic fields, and one way of showing that a solution contains free radicals is to show that it is attracted by a magnet. Unfortunately, this method is not very sensitive, and only relatively large concentrations of radicals can be detected. Much more sensitive is the method of *electron spin resonance spectroscopy,* often called e.s.r. (or e.p.r., electron paramagnetic resonance) for short. This depends on the fact that a free radical in a strong magnetic field will preferentially orient its odd electron spin in the more stable direction (as a compass needle aligns in the earth's field) but with the absorption of light energy (with the usual magnetic fields, microwave frequencies are used) the spin can turn over to the unstable orientation, as illustrated in Figure 7–1.

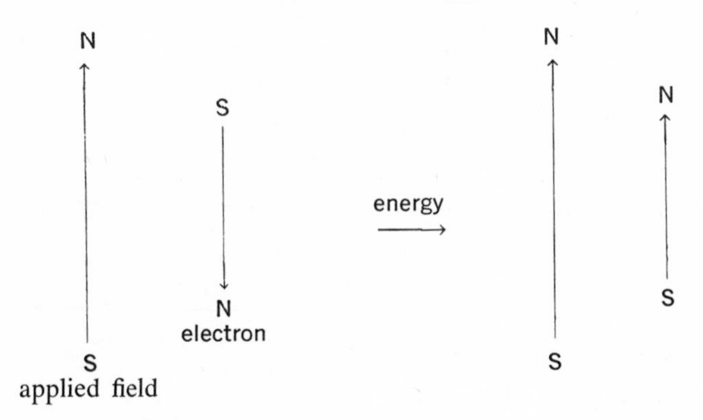

FIGURE 7–1

Illustration of the principle of e.s.r.

This is a very sensitive method of detecting radicals, even at concentrations of $10^{-7}M$.

Using e.s.r. it has been possible to detect the intermediate radicals in a free radical polymerization. Furthermore, even very unstable radicals such as $CH_3\cdot$ can be detected using e.s.r.; conveniently such reactive radicals are trapped at −200°C in a frozen solution so that they cannot react and be destroyed. Using e.s.r., many radicals which were formerly only postulated intermediates have now been directly observed.

7–2 Bond Dissociation Energies

For radicals which are too unstable to exist at equilibrium, information on relative stabilities can be obtained from bond dissociation energies (the energy required to break a covalent bond and form two radicals). Table 7–1 contains values for some molecules of interest; these are determined in a few cases by direct spectroscopic measurements, but in most instances by more complex indirect means.

It is apparent that the relative stabilities of different radicals cannot be derived from this table without the exercise of some care. Thus F_2 has a low dissociation energy compared with Cl_2,

TABLE 7–1 ▪

Some Bond Dissociation Energies (Kcal/Mole at 25°C)

Bond	Dissociation Energy	Bond	Dissociation Energy
CH_3—H	102	H—H	104
CH_3CH_2—H	98	Cl—Cl	58
$(CH_3)_2CH$—H	94	Br—Br	46
$(CH_3)_3C$—H	90	F—F	37
$C_6H_5CH_2$—H	78	I—I	36
C_6H_5—H	102	HO—OH	52
CCl_3—H	90	O_2N—NO_2	13
O = CH—H	78	H_2N—NH_2	60
HO—H	120	CH_3—CH_3	84
HOO—H	90	CH_3—OH	90
F—H	135	CH_3—Cl	82
Cl—H	103	CH_3—Br	67
Br—H	87	CH_3—I	53
I—H	71		
CH_3S—H	89		

suggesting that the fluorine radical (atom) is more stable than the chlorine atom. On the other hand, HF has a higher dissociation energy than does HCl, suggesting that a chlorine atom is more stable than a fluorine. The reason for this dilemma is that we have ignored the properties of the bond which is being broken; the single bond in HF is much more stable than the single bond in F_2, while the HCl/Cl—Cl difference is smaller. Bond dissociation energies can be used as an indication of radical stabilities only if a very similar bond is being broken in all the cases compared. This is approximately true for the first five hydrocarbons in the table, and to a lesser extent for the next three cases as well in which a C—H bond is being cleaved. Thus the values in the table show that tertiary alkyl radicals are more stable than secondary, which are more stable than primary. The stabilizing effect of alkyl substituents is often ascribed to hyperconjugation.

$$(CH_3)_2\dot{C}\text{—}CH_3 \longleftrightarrow (CH_3)_2C{=}CH_2\ \cdot H \longleftrightarrow (CH_3)_2C{=}CH_2\ \cdot H \quad \text{etc.}$$

The benzyl radical is strongly stabilized by conjugation of the odd electron with the benzene ring; trichloromethyl radical and formyl radical are also stabilized by conjugation.

etc.

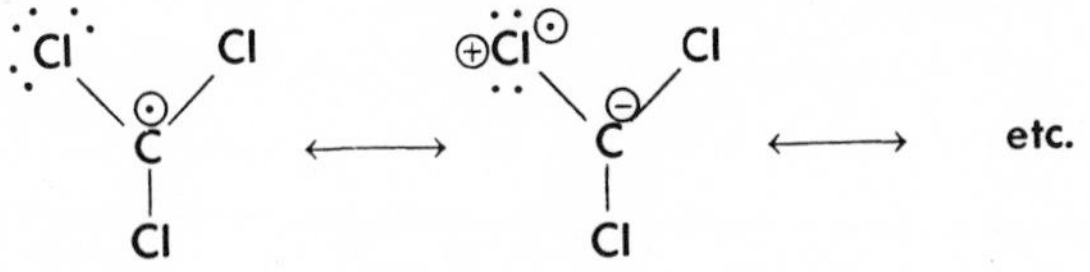

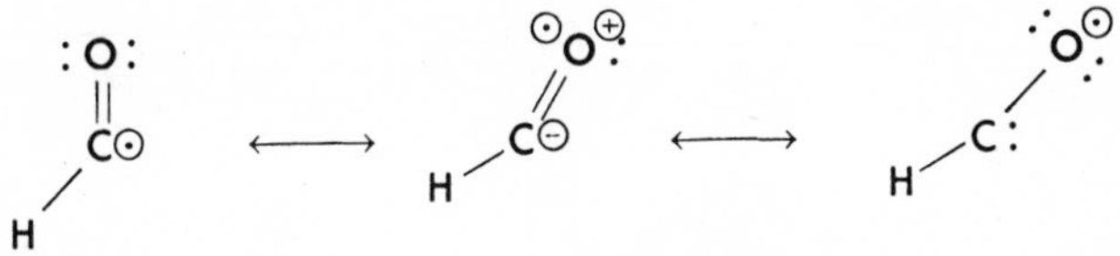

The same sort of stabilization is found in hydroperoxy radical, HOO·, compared with hydroxyl radical.

Bond dissociation energies can be used to predict whether various simple radical reactions are exothermic or not. Thus we see that hydroxyl radical can attack methane in an exothermic process, but that a bromine atom cannot.

$$HO\cdot + CH_4 \rightarrow H_2O + CH_3\cdot \quad \Delta H^\circ = -18 \text{ kcal/mole}$$

$$Br\cdot + CH_4 \rightarrow HBr + CH_3\cdot \quad \Delta H^\circ = +15 \text{ kcal/mole}$$

The over-all energy change is just the difference of two bond dissociation energies. Breaking a methyl-hydrogen bond requires 102 kcal/mole, but forming a hydroxyl-hydrogen bond releases 120, so the over-all process releases 18 kcal/mole.

Since the reaction of a bromine atom with methane is endothermic, the reverse reaction—attack on HBr by a methyl radical—is exothermic. There are actually two ways in which $CH_3\cdot$ could attack HBr, to form methane or to form methyl bromide.

$$CH_3\cdot + HBr \rightarrow CH_4 + Br\cdot \quad \Delta H^\circ = -15 \text{ kcal/mole}$$

$$CH_3\cdot + HBr \rightarrow CH_3Br + H\cdot \quad \Delta H^\circ = +20 \text{ kcal/mole}$$

The difference between a bond dissociation energy of 67 kcal for methyl bromide and 87 kcal for HBr shows that the second reaction is endothermic by 20 kcal/mole and the first reaction is thus the favored one. This is very important, for reactions which are strongly endothermic cannot play a major role in radical chain reactions. Chain reactions are the most important feature of free radical chemistry.

7–3 Radical Chain Reactions

Polymerization. The free radical polymerization of styrene, initiated by benzoyl peroxide, is a typical chain reaction.

$$C_6H_5\overset{\overset{O}{\|}}{C}O-O\overset{\overset{O}{\|}}{C}C_6H_5 \xrightarrow{\text{Heat}} 2\ C_6H_5\overset{\overset{O}{\|}}{C}-O\cdot$$

$$C_6H_5\overset{\overset{O}{\|}}{C}-O\cdot + CH_2{=}CHC_6H_5 \longrightarrow C_6H_5\overset{\overset{O}{\|}}{C}-OCH_2\dot{C}H-C_6H_5$$

$$R\cdot + CH_2{=}CHC_6H_5 \longrightarrow RCH_2\dot{C}HC_6H_5$$

$$2\,R\cdot \longrightarrow R-R$$

The first step is called the *chain initiation step*. Here a special substance is used which has a relatively low bond dissociation energy; cleavage occurs not only because a weak oxygen-oxygen bond is being broken (cf. hydrogen peroxide in Table 7–1) but also because the product radicals are resonance-stabilized.

$$C_6H_5-C(=O)-O^{\bullet} \longleftrightarrow C_6H_5-C(-{}^{\bullet}O)=O$$

In step 2 this radical adds to the styrene double bond. This is a favorable step because the product radical is also resonance stabilized, and because a relatively strong carbon-oxygen single bond (cf. CH_3—OH in Table 7–1) is being formed while a relatively weak carbon-carbon π bond is being broken.

Step 3 is the chain-propagating step. It occurs over and over again, the radical becoming longer with each addition to a styrene molecule. This step is favorable because a new carbon-carbon single bond is formed while a carbon-carbon double bond is broken; the single bond is about 20 kcal/mole stronger than a π bond.

Eventually two of the growing radical chains collide and react. Then *chain termination,* step 4, occurs.

$$C_6H_5\overset{O}{\overset{\|}{C}}O(CH_2\overset{C_6H_5}{\overset{|}{C}H})_nCH_2\overset{C_6H_5}{\overset{|}{C}H}\cdot + \cdot\overset{C_6H_5}{\overset{|}{C}H}CH_2(\overset{C_6H_5}{\overset{|}{C}H}CH_2)_mO\overset{O}{\overset{\|}{C}}C_6H_5 \longrightarrow$$

$$\diagdown CH_2\overset{C_6H_5}{\overset{|}{C}H}—\overset{C_6H_5}{\overset{|}{C}H}\diagup$$

Termination has been written as coupling of the two radicals, but it can also occur by hydrogen transfer.

$$\diagdown CH_2\overset{C_6H_5}{\overset{|}{C}H}\cdot + \cdot\overset{C_6H_5}{\overset{|}{C}H}CH_2\diagup \longrightarrow \diagdown CH_2\overset{C_6H_5}{\overset{|}{C}H_2} + \overset{C_6H_5}{\overset{|}{C}H}=CH—$$

There is a high probability that two such radicals will react if they collide, and in fact it has been found that two methyl radicals will couple to form ethane on almost every collision. Thus it will be possible to obtain long-chain polymers only if the chain propagation step is also very fast. Typically, the time between chain initiation and chain termination is of the order of one second. During this time all of the chain-propagating steps occur.

We have already emphasized the fact that reactions which are strongly endothermic cannot play an important role as chain-propagating steps. This is because a reaction must have a low activation energy in order to be fast, and the activation energy for any step can never be less than the over-all energy change for that step. This is illustrated in the energy diagram of Figure 7–2.

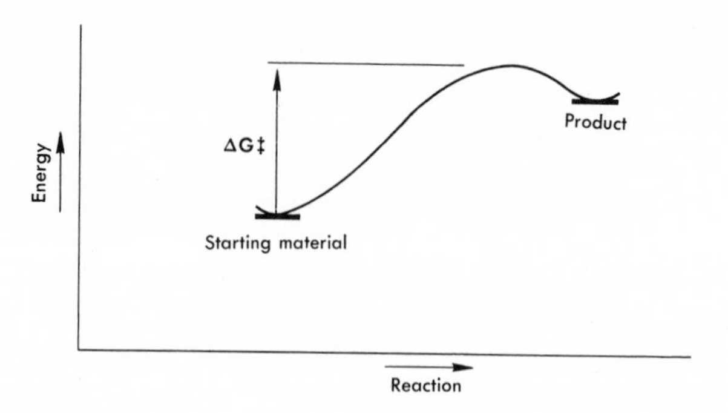

FIGURE 7–2
Energy diagram for an endothermic reaction.

It is apparent that the transition state must be at least as high in energy as the products are, since the transition state is the highest energy point on the path. When we recall that at room temperature each 1.4 kcal/mole of extra activation energy slows a reaction by a factor of 10, it is easy to see that a reaction which is endothermic by 20 kcal/mole, for instance, will be very slow indeed, too slow to play a role as a chain-propagating step. Although even an exothermic process could in principle have a large activation energy, the exothermic addition of a radical to another styrene unit is actually relatively fast, and long chains are produced.

Many other olefins can be polymerized by radical chain processes, including ethylene, vinyl acetate, vinyl chloride, acrylonitrile, etc. In general a reactive monomer will have a terminal double bond and addition occurs at the unhindered CH_2 to yield the more stable substituted radical.

Substitution. Molecular fluorine is enormously reactive towards hydrocarbons. The careful reaction of F_2 with methane yields methyl fluoride, along with further fluorination products. For this reaction the chain propagation steps are the following:

$$F\cdot + CH_4 \rightarrow HF + CH_3\cdot \quad \Delta H^\circ = -33 \text{ kcal/mole}$$

$$CH_3\cdot + F_2 \rightarrow CH_3F + F\cdot \quad \Delta H^\circ = -60 \text{ kcal/mole}$$

Both steps are very exothermic because the hydrogen-fluorine and carbon-fluorine bond are strong while the F_2 bond is weak.

The initiation step is probably a direct reaction of methane with fluorine, since no other initiators are required.

$$CH_4 + F_2 \rightarrow CH_3\cdot + HF + F\cdot \qquad \Delta H° = +4 \text{ kcal/mole}$$

Chain termination comes by combination of two methyl radicals, by combination of methyl with fluorine radicals, and perhaps by recombination of fluorine atoms to form F_2.

Molecular chlorine is much less reactive, although it can give a chain process in which both steps are exothermic.

$$Cl\cdot + CH_4 \rightarrow HCl + CH_3\cdot \qquad \Delta H° = -1 \text{ kcal/mole}$$

$$CH_3\cdot + Cl_2 \rightarrow CH_3Cl + Cl\cdot \qquad \Delta H° = -24 \text{ kcal/mole}$$

These energies can be obtained from Table 7–1 by a combination of the appropriate bond dissociation energies. For instance, for the reaction of methyl radical with Cl_2 the energy change is just the difference between the 58 kcal of the Cl_2 bond and the 82 kcal of the methyl chloride bond. Again using the data in this table, it is apparent that a direct reaction between Cl_2 and methane is probably not the initiating step, since it would be endothermic by 57 kcal. Accordingly, free radical chlorinations are generally initiated with light, although other initiators may also be used.

$$Cl_2 \xrightarrow{\text{light}} 2\,Cl\cdot$$

The 58 kcal/mole required to break this bond is furnished by the light which is absorbed. The energy of a quantum of light depends on the wavelength, of course, and it can be derived from a simple formula:

$$E \text{ (in kcal/mole)} = \frac{28{,}635}{\lambda \text{ (in m}\mu)}$$

Thus blue light at 400 mμ has an energy of 71.6 kcal/mole, while ultraviolet light at 200 mμ can furnish 143.2 kcal/mole. This energy is enough to break any covalent bond, if it is absorbed; therefore ultraviolet irradiation is often used to initiate radical chain reactions.

Free radical bromination is even less easy than chlorination. Since the HBr bond has a strength of only 87 kcal, Br· can attack only fairly weak C—H bonds. We have already seen that attack on methane would be endothermic by 15 kcal; on the other hand, the hydrogen of toluene can be abstracted in an exothermic process (cf. Table 7–1).

$$Br\cdot + H{-}CH_2C_6H_5 \rightarrow HBr + C_6H_5CH_2\cdot \qquad \Delta H^\circ = -9 \text{ kcal}$$

$$C_6H_5CH_2\cdot + Br_2 \rightarrow C_6H_5CH_2Br + Br\cdot \qquad \Delta H^\circ = -4 \text{ kcal}$$

Brominations of this type are usually initiated either with light or by the thermal decomposition of a peroxide or other unstable molecule.

Since an allyl radical is about as stable as a benzyl radical, it would be expected that hydrogens next to a double bond could also be replaced in a chain bromination. This is true, but olefins can also add bromine to the double bond in a polar reaction. To prevent this polar addition, *N*-bromosuccinimide is usually used to effect allylic bromination by a free radical process.

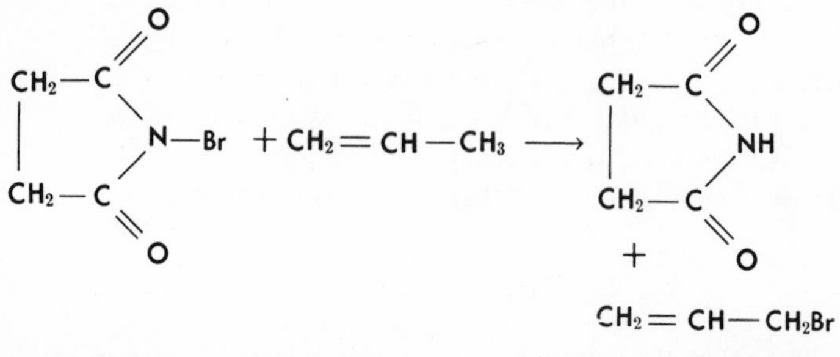

$$(CH_2CO)_2N{-}Br + CH_2{=}CH{-}CH_3 \longrightarrow (CH_2CO)_2NH + CH_2{=}CH{-}CH_2Br$$

Although it was once thought that this involved a special pathway, it is now clear that the reaction of *N*-bromosuccinimide with HBr formed in side reactions simply furnishes Br_2 in *low concentrations*. Under these conditions addition of bromine to the double bond is slow, and free radical bromination occurs instead.

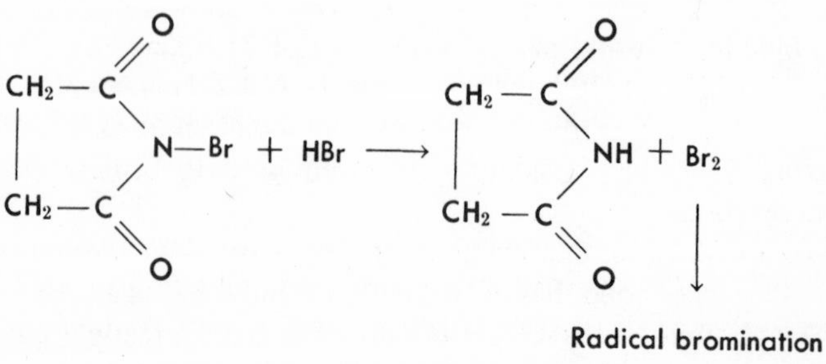

$$(CH_2CO)_2N{-}Br + HBr \longrightarrow (CH_2CO)_2NH + Br_2$$

$$Br_2 \downarrow \text{ Radical bromination}$$

Free radical halogenation with iodine is generally not possible, since the 71 kcal bond energy of HI is lower than the energy of any C—H bond.

The combustion of hydrocarbons involves a free radical chain substitution by O_2, followed by further complex changes. The first step may be illustrated by the *autoxidation* (i.e., reaction with molecular oxygen) of benzaldehyde.

$$R\cdot + C_6H_5C(=O)H \longrightarrow RH + C_6H_5-\dot{C}=O$$

$$C_6H_5-\dot{C}=O + O_2 \longrightarrow C_6H_5-C(=O)-O-O\cdot$$

$$C_6H_5-C(=O)-O-O\cdot + C_6H_5C(=O)H \longrightarrow C_6H_5-C(=O)-O-OH + C_6H_5\dot{C}=O$$

A radical formed in an initiating step removes the very reactive (cf. Table 7–1) aldehyde hydrogen, and the next two steps are then chain-propagating. The product of this chain is perbenzoic acid, but this reacts with benzaldehyde in an ionic process to afford benzoic acid.

$$C_6H_5C(=O)-O-OH + C_6H_5C(=O)H \longrightarrow 2\ C_6H_5C(=O)OH$$

Generally the step in which a radical adds to O_2 is quite favorable energetically, but in the second propagation step a relatively weak (90 kcal) ROO—H bond is being formed. Accordingly autoxidation at ordinary temperatures is facile only with relatively reactive C—H bonds, such as those in aldehydes or the C—H bond next to an ether oxygen. Allylic hydrogens are also susceptible to autoxidation; this is part of the reason that unsaturated fatty acid esters turn rancid in air.

Addition. The most famous free radical addition process is the "abnormal" addition of HBr to olefins. With propylene, for instance, the chain-propagating steps are as follows:

$$Br\cdot + CH_2{=}CH{-}CH_3 \longrightarrow Br{-}CH_2{-}\dot{C}H{-}CH_3 \qquad \Delta H° = -9\ \text{kcal/mole}$$

$$Br{-}CH_2{-}\dot{C}H{-}CH_3 + HBr \longrightarrow Br{-}CH_2{-}CH_2{-}CH_3 + Br\cdot \qquad \Delta H° = -7\ \text{kcal/mole}$$

The over-all result is "anti-Markownikoff" addition, i.e., the reverse of the orientation in ionic addition to an olefin; the orientation is as expected for a radical chain since bromine adds to the less hindered end of the bond to form at the same time the more stable radical. Chain initiation may be effected with light or by the thermal decomposition of initiators such as peroxides. Indeed, the peroxides present in most olefin samples, because of autoxidation, are usually sufficient to initiate these chains; olefins must be carefully purified if the radical process is to be suppressed.

$$ROOH \xrightarrow[\text{light}]{\text{heat or}} RO\cdot + \cdot OH$$

$$RO\cdot + HBr \rightarrow ROH + Br\cdot$$

Chain termination occurs by combination of bromine atoms with themselves or with alkyl radicals, or by coupling or disproportionation of alkyl radicals.

It is interesting that polymer is never formed under these conditions. In principle the alkyl radical could add to another olefin molecule, as in olefin polymerization, but in practice with any reasonable concentration of HBr the latter traps all alkyl radicals formed before they can attack a second olefin molecule.

Neither HI nor HF has so far (1969) been added to an olefin in a free radical chain process. To understand this we must simply consider the energy changes which would be associated with each propagating step of the chain. Thus for the HI addition to ethylene

$$I\cdot + CH_2 = CH_2 \rightarrow I{-}CH_2{-}CH_2\cdot \qquad \Delta H^\circ = +7 \text{ kcal/mole}$$

$$I{-}CH_2{-}CH_2\cdot + HI \rightarrow I{-}CH_2{-}CH_2{-}H + I\cdot \qquad \Delta H^\circ = -27 \text{ kcal/mole}$$

The first step is unfavorable because the carbon-iodine bond is so weak (with substituted olefins in which a highly stabilized radical is formed this step might become favorable). The second step is fine, but with one slow step the chain propagation cannot compete with termination. For HF the slow step is the second one.

$$F{-}CH_2{-}CH_2\cdot + HF \rightarrow F{-}CH_2{-}CH_2{-}H + F\cdot \qquad \Delta H^\circ = +37 \text{ kcal/mole}$$

This value is obtained from Table 7–1 simply by considering the bond strengths in ethane and in HF.

In general, HCl also fails to add to olefins in a radical chain process. Addition of a chlorine atom to ethylene is exothermic by 26 kcal/mole, but the reaction of the product radical with HCl is unfavorable by 5 kcal/mole, as the bond energies of ethane and of HCl in Table 7–1 show. Slow addition of HCl in a radical

$$C_6H_5\overset{O}{\overset{\|}{C}}O{-}O\overset{O}{\overset{\|}{C}}C_6H_5 \longrightarrow 2\ C_6H_5\overset{O}{\overset{\|}{C}}{-}O\cdot$$

$$C_6H_5C(=O)O\cdot + HCl \longrightarrow C_6H_5C(=O)OH + Cl\cdot$$

$$Cl\cdot + CH_2{=}CH_2 \longrightarrow Cl{-}CH_2{-}CH_2\cdot$$

$$Cl{-}CH_2CH_2\cdot + CH_2{=}CH_2 \longrightarrow Cl(CH_2CH_2)_2\cdot \qquad \text{etc.}$$

$$Cl{-}(CH_2CH_2)_n\cdot + HCl \longrightarrow Cl(CH_2CH_2)_nH + Cl\cdot \qquad n = 2 \text{ to } > 10$$

process is possible with some olefins, but the slowness of the second step means that polymerization now competes with simple addition. Thus short polymers (called "telomers") are obtained when ethylene is heated under pressure with aqueous HCl, using benzoyl peroxide as initiator.

Thiols of all sorts can be added to double bonds in free radical chain reactions. For instance, thiophenol reacts with styrene to afford a simple adduct.

$$C_6H_5S\cdot + CH_2{=}CH-C_6H_5 \longrightarrow C_6H_5S-CH_2\dot{C}H-C_6H_5$$

$$C_6H_5S-CH_2-\dot{C}H-C_6H_5 + C_6H_5SH \longrightarrow C_6H_5SCH_2CH_2C_6H_5 + C_6H_5S\cdot$$

Both steps are exothermic, and polymers are not formed since the hydrogen transfer step is so rapid.

Carbon-carbon bonds can also be formed in some free radical addition reactions. For instance, carbon tetrachloride can be added to propylene in 80% yield.

$$\cdot CCl_3 + CH_2{=}CH-CH_3 \longrightarrow Cl_3C-CH_2-\dot{C}H-CH_3$$

$$Cl_3C-CH_2-\dot{C}H-CH_3 + CCl_4 \longrightarrow Cl_3C-CH_2\underset{|}{\overset{Cl}{C}}H-CH_3 + \cdot CCl_3$$

The reaction is initiated by the thermal decomposition of dibenzoylperoxide; the benzoate radicals attack CCl_4 to produce a $CCl_3\cdot$ and start the chain. The first propagation step is favorable because a carbon-carbon single bond replaces a double bond. This overcomes the fact that the radical formed is less stable than the trichloromethyl radical consumed (cf. chloroform and propane in Table 7–1). In the second step there is no change in the types of bonds present, but a conversion of the alkyl radical to a more stable trichloromethyl radical, so it is favorable as well.

Aldehydes can also be added to some olefins. Thus reaction of acetaldehyde with diethyl maleate produces the adduct in 78% yield.

$$CH_3\overset{O}{\overset{\|}{C}}\cdot + \underset{C_2H_5O_2C}{\overset{H}{C}}{=}\underset{CO_2C_2H_5}{\overset{H}{C}} \longrightarrow CH_3{-}\overset{O}{\overset{\|}{C}}{-}\underset{C_2H_5O_2C}{\overset{H}{C}}{-}\underset{CO_2C_2H_5}{\dot{C}H}$$

$$\xrightarrow{CH_3\overset{O}{\overset{\|}{C}}{-}H} CH_3{-}\overset{O}{\overset{\|}{C}}{-}\underset{C_2H_5O_2C}{CH}{-}\underset{CO_2C_2H_5}{CH_2} + CH_3\overset{O}{\overset{\|}{C}}\cdot$$

Even alcohols can be added, ethanol reacting with ethylene to afford a 10% yield of the simple adduct.

$$CH_3\underset{\bullet}{\overset{OH}{C}}H + CH_2{=}CH_2 \longrightarrow CH_3{-}\overset{OH}{\overset{|}{C}H}{-}CH_2{-}CH_2\cdot \xrightarrow{CH_3CH_2OH}$$

$$CH_3\overset{OH}{\overset{|}{C}H}{-}CH_2{-}CH_3 + CH_3\overset{OH}{C}H\cdot$$

In many of these cases telomers, short polymers, are formed as well.

Several features will be noted which are common to all these mechanisms. After initiation steps, which may occur in various ways depending on the reaction considered, the two steps of chain propagation consist of (1) *addition* to a double bond, and (2) free radical *substitution.* By contrast, polymerization consists of a series of *addition* reactions, and halogenation of hydrocarbons consisted of a chain of alternating *substitution* reactions. The other significant point is that in general addition to double bonds occurs at their least hindered end, and radical substitution involves attack on unhindered atoms like hydrogen or halogen on the outside of a molecule, rather than attack on a carbon atom inside a molecule. The requirement that chain-propagating steps have a fairly rapid rate means not only that they should be exothermic, but also that there should be no serious steric barrier to be overcome.

Free radical decompositions and rearrangements. So far we have considered only the ways in which free radicals may attack other molecules. In some cases intramolecular reactions of radicals may also be important. For instance, we have frequently referred to the use of benzoyl peroxide as a radical initiator, and have suggested that it functions by furnishing benzoate radicals. In fact, this is only part of the story. Decomposition of benzoyl peroxide by heating in cyclohexane affords both benzoic acid and benzene, together with other products derived from the benzoate and phenyl radicals.

$$C_6H_5\overset{\overset{\displaystyle O}{\|}}{C}{-}O{-}O{-}\overset{\overset{\displaystyle O}{\|}}{C}C_6H_5 \xrightarrow{\Delta} 2\ C_6H_5\overset{\overset{\displaystyle O}{\|}}{C}{\diagdown}O\cdot$$

$$C_6H_5\overset{\overset{\displaystyle O}{\|}}{C}{\diagdown}O\cdot + RH \longrightarrow C_6H_5CO_2H + R\cdot$$

$$C_6H_5\overset{\overset{\displaystyle O}{\|}}{C}{\diagdown}O\cdot \longrightarrow C_6H_5\cdot + CO_2$$

$$C_6H_5\cdot + RH \longrightarrow C_6H_6 + R\cdot$$

Carbon dioxide is also found, as expected since the benzoate radical decarboxylates to form a phenyl radical.

Similarly acetyl peroxide decomposes, at 60–100°C, to yield methyl radicals almost exclusively; decomposition of the intermediate acetate radical occurs before it can be trapped.

$$CH_3CO{-}O{-}O{-}COCH_3 \longrightarrow CH_3CO{-}O\cdot \longrightarrow \cdot CH_3 + CO_2$$

On standing at room temperature, trimethylacetaldehyde decomposes to carbon monoxide and isobutane.

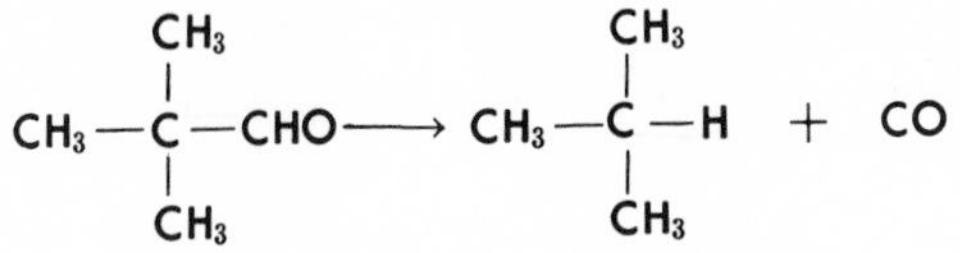

$$\begin{array}{c} CH_3 \\ | \\ CH_3-C-CHO \\ | \\ CH_3 \end{array} \longrightarrow \begin{array}{c} CH_3 \\ | \\ CH_3-C-H \\ | \\ CH_3 \end{array} + CO$$

This is a free radical chain process, initiated presumably by traces of peroxides in the aldehyde.

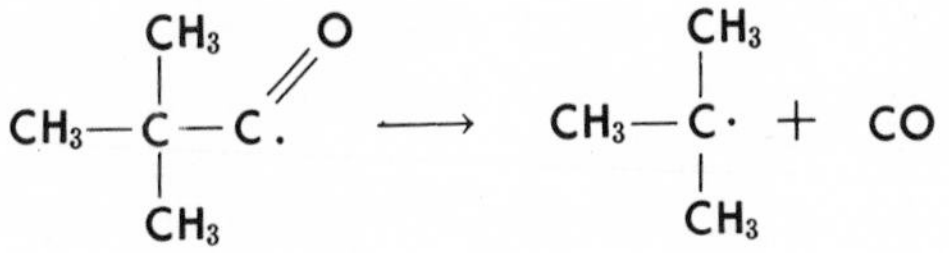

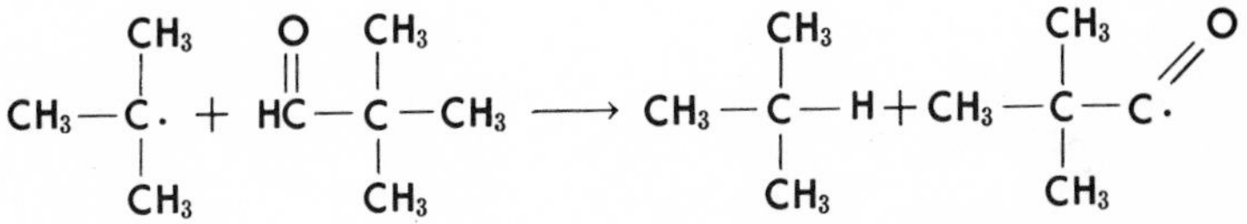

$$\begin{array}{c} CH_3\quad O \\ |\quad // \\ CH_3-C-C\cdot \\ | \\ CH_3 \end{array} \longrightarrow \begin{array}{c} CH_3 \\ | \\ CH_3-C\cdot \\ | \\ CH_3 \end{array} + CO$$

$$\begin{array}{c} CH_3 \\ | \\ CH_3-C\cdot \\ | \\ CH_3 \end{array} + \begin{array}{c} O\quad CH_3 \\ \|\quad | \\ HC-C-CH_3 \\ | \\ CH_3 \end{array} \longrightarrow \begin{array}{c} CH_3 \\ | \\ CH_3-C-H \\ | \\ CH_3 \end{array} + \begin{array}{c} CH_3\quad O \\ |\quad // \\ CH_3-C-C\cdot \\ | \\ CH_3 \end{array}$$

Although this is a particularly easy decarbonylation, many acyl radicals lose CO on heating, and such decarbonylations represent attractive ways to generate alkyl radicals. For instance, heating trimethylacetaldehyde with 1-hexene leads to 2,2-dimethyloctane in good yield.

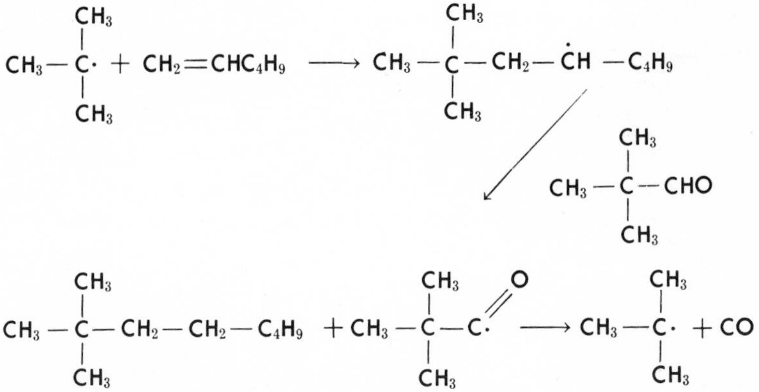

$$\begin{array}{c} CH_3 \\ | \\ CH_3-C\cdot \\ | \\ CH_3 \end{array} + CH_2{=}CHC_4H_9 \longrightarrow \begin{array}{c} CH_3 \\ | \\ CH_3-C-CH_2-\dot{C}H-C_4H_9 \\ | \\ CH_3 \end{array}$$

$$\swarrow \begin{array}{c} CH_3 \\ | \\ CH_3-C-CHO \\ | \\ CH_3 \end{array}$$

$$\begin{array}{c} CH_3 \\ | \\ CH_3-C-CH_2-CH_2-C_4H_9 \\ | \\ CH_3 \end{array} + \begin{array}{c} CH_3\quad O \\ |\quad // \\ CH_3-C-C\cdot \\ | \\ CH_3 \end{array} \longrightarrow \begin{array}{c} CH_3 \\ | \\ CH_3-C\cdot \\ | \\ CH_3 \end{array} + CO$$

Fragmentation also occurs when di-*t*-butyl peroxide is heated, the *t*-butoxy radical undergoing some cleavage to acetone and a methyl radical.

$$(CH_3)_3C{-}O{-}O{-}C(CH_3)_3 \xrightarrow{\text{Heat}} (CH_3)_3C{-}O\cdot \longrightarrow CH_3\cdot + (CH_3)_2C{=}O$$

Thus in the presence of solvents which can donate hydrogen to a radical, both *t*-butyl alcohol and methane are formed.

Finally, some mention should be made of the rearrangements of free radicals. These are relatively rare. Generally one does not observe shifts of alkyl groups or of hydrogen atoms in free radicals, in contrast to the situation with carbonium ions, but the migration of phenyl groups is common. In the (unknown) migration of a methyl group, an intermediate configuration would have to be that shown below.

$$CH_3{-}CH_2{-}\dot{C}H_2 \longleftrightarrow \dot{C}H_2{-}CH_2{-}CH_3 \equiv \text{bridged } CH_2\cdots CH_3 \cdots CH_2 \text{ (dotted-line intermediate)}$$

This is just like the intermediate geometry in a carbonium ion rearrangement, but for the radical there are three electrons spread over the three carbon atoms while in a carbonium ion rearrangement there are only two. The situation turns out to be very similar to that discussed in Special Topic 1, where it was seen that the cyclopropenyl cation, with two π electrons spread over three carbons, was much more stable than a cyclopropenyl radical with three π electrons. Detailed molecular orbital calculations on the intermediates for carbonium ion and radical rearrangements show here as well that the three electron case is much less favorable. On the other hand, phenyl migration can occur by bridged radical formation, without the need for "dotted line" intermediates.

$$C_6H_5{-}CH_2{-}\dot{C}H_2 \longrightarrow \text{spiro-cyclopropyl cyclohexadienyl radical} \longrightarrow \dot{C}H_2{-}CH_2{-}C_6H_5$$

This exact rearrangement has been detected, using C^{14} as a tracer, but only 4% of the product was derived from phenyl migration.

$$C_6H_5CH_2\overset{*}{C}H_2CHO \xrightarrow{R\cdot} C_6H_5CH_2\overset{*}{C}H_2\overset{O}{\overset{\|}{C}}\cdot \xrightarrow[C]{165^\circ} C_6H_5CH_2\overset{*}{C}H_2\cdot \longrightarrow \cdot CH_2-\overset{*}{C}H_2C_6H_5$$

$$C_6H_5CH_2\overset{*}{C}H_2\cdot \xrightarrow{RCHO} C_6H_5CH_2\overset{*}{C}H_3 \quad 96\%$$

$$\cdot CH_2-\overset{*}{C}H_2C_6H_5 \xrightarrow{RCHO} CH_3\overset{*}{C}H_2C_6H_5 \quad 4\%$$

However, the decarbonylation of β-phenylisovaleraldehyde affords a mixture of hydrocarbons including as much as 80% of the product from phenyl migration.

$$C_6H_5-\underset{CH_3}{\overset{CH_3}{\overset{|}{\underset{|}{C}}}}-CH_2CHO \xrightarrow[\text{Initiator}]{\text{Heat}} \underset{20\%}{C_6H_5\underset{CH_3}{\overset{CH_3}{\overset{|}{\underset{|}{C}}}}CH_3} + \underset{80\%}{C_6H_5CH_2-\underset{CH_3}{\overset{CH_3}{\overset{|}{\underset{|}{C}}}}H} + CO$$

General References

W. Pryor, *Free Radicals* (McGraw-Hill, New York, 1966). The best general introduction, with a good list of references.

C. Walling, *Free Radicals in Solution* (John Wiley & Sons, New York, 1957). The classic book in this field.

C. Walling and E. Huyser, "Free Radical Additions to Olefins to Form Carbon-Carbon Bonds" and F. W. Stacey and J. F. Harris, "Formation of Carbon-Hetero Atom Bonds by Free Radical Chain Additions to Carbon-Carbon Multiple Bonds," in A. Cope, ed., *Organic Reactions,* Vol. XIII (John Wiley & Sons, New York, 1963). Guides to more recent work.

J. A. McMillan, "Electron Paramagnetic Resonance of Free Radicals," *Journal of Chemical Education,* **38,** 438 (1961).

B. A. Bohm and P. I. Abell, "Stereochemistry of Free Radical Additions to Olefins," *Chemical Reviews,* **62,** 599 (1962). The stereochemistry of free radical reactions is still under active investigation.

O. Dermer and M. Edmison, "Radical Substitution in Aromatic Nuclei," *Chemical Reviews*, **57**, 77 (1957). An interesting topic which has not been covered in this book.

H. Lankamp, W. Th. Nauta, and C. MacLean, "A New Interpretation of the Monomer-Dimer Equilibrium of Triphenylmethyl and Alkyl Substituted-Diphenylmethyl Radicals in Solution," *Tetrahedron Letters*, **249** (1968). Modern tools show that "hexaphenylethane" has another structure!

7

Special Topic

▪ POLAR EFFECTS IN FREE RADICAL REACTIONS

THROUGHOUT THE LAST CHAPTER we have emphasized the fact that chain-propagating steps must be very rapid. Even so, it is apparent that these reactions do not in general occur with every collision, for radical reactions can often be quite selective. In considering the relative rate at which two competing free radical processes might occur, e.g., addition at one or the other end of the styrene double bond, we have explained the preference for addition at the unsubstituted end in terms of *steric factors* and in terms of *product stability,* it being implied that the transition state for

$$R\cdot + CH_2{=}CH{-}C_6H_5 \longrightarrow R{-}CH_2{-}\dot{C}H{-}C_6H_5$$

$$R\cdot + CH_2{=}CH{-}C_6H_5 \not\longrightarrow R{-}CH(C_6H_5){-}CH_2\cdot$$

addition at least partly resembles the product. There is now increasing evidence that *polar effects* also often play a role in determining the stability of such a transition state.

One of the clearest examples of this is found in free radical bromination of the methyl group in substituted toluenes. In Table 7–2 are listed relative rates, determined with Br_2 in benzene solution at 80°C.[1]

TABLE 7–2 ■

Relative Rates of Free Radical Bromination of Substituted Toluenes[1]

Substituent	Relative Rate	σ^+
None	1.00	0
p-OCH_3	9.00	− 0.778
p-CH_3	2.42	− 0.256
p-*t*-Butyl	2.47	− 0.311
p-Cl	0.73	+ 0.405
m-Br	0.22	+ 0.114
p-CN	0.12	+ 0.659

It is not at all surprising that a *p*-methoxy group activates toluene towards hydrogen abstraction, for the product radical has additional stability because of resonance forms which involve the oxygen.

CH_3–C_6H_4–$\ddot{O}CH_3$ + Br· ⟶ HBr + $\dot{C}H_2$–C_6H_4–$\ddot{O}CH_3$ ⟷ CH_2=C_6H_4(⊙)–$\ddot{O}CH_3$ ⟷ CH_2=C_6H_4(⊖)–⊕⊙$\ddot{O}CH_3$

Similarly a *p*-methyl stabilizes the product radical, and thus also stabilizes the transition state to the extent that this resembles the product. However, a *p*-cyano group deactivates the toluene, although it can also conjugate with the radical.

1. R. Pearson and J. Martin, "The Mechanism of Benzylic Bromination," *Journal of the American Chemical Society*, **85,** 354 (1963).

$$p\text{-}\dot{C}H_2C_6H_4C\equiv N \longleftrightarrow CH_2{=}C_6H_4{=}C{=}\dot{N}$$

Furthermore, a *m*-bromo substituent is also strongly deactivating. This result cannot be explained in terms of stability of the product radical, which should be unaffected by this unconjugated group.

It seems apparent that the reaction is favored by groups which can stabilize a positive charge (i.e., those with a negative σ^+) and slowed by those which destabilize a positive charge. This polar effect can be explained[2] if one considers the structure of the transition state for hydrogen abstraction.

$$C_6H_5CH_2\text{---}H\text{---}Br \equiv \underset{\text{I}}{C_6H_5CH_2{-}H \cdot Br} \longleftrightarrow \underset{\text{II}}{C_6H_5CH_2\cdot \; H{-}Br} \longleftrightarrow \underset{\text{III}}{C_6H_5CH_2^{\oplus} \; \dot{H} \; Br^{\ominus}}$$

As the bromine atom begins to bond to the hydrogen, a species is formed in which three electrons are distributed over two partial bonds, shown in dotted lines. This species can also be represented as a hybrid of three resonance forms: one in which there is a carbon-hydrogen "long bond" (I), one in which there is instead a bromine-hydrogen "long bond" (II), and the third ionic form (III). Here it is recognized that since bromine is more electronegative than carbon there will be a drift of electrons from left to right, so carbon will acquire some positive character while bromine gets some negative charge. To the extent that the transition state resembles resonance form II, all conjugating substituents should help; to the extent that form III plays a role, then polar effects will be seen.

2 (a). C. Walling, *Free Radicals in Solution* (John Wiley & Sons, New York, 1957) (b). W. Pryor, *Free Radicals* (McGraw-Hill Book Co., New York, 1966). Cf. the index under "Polar Effects."

A similar polar effect is seen[3] in the chlorination of *n*-butyl chloride (initiated by light). The relative reactivities of the hydrogens on the four carbons are as indicated below.

$$\begin{array}{ccccccccc} 5.6 & & 17 & & 5.6 & & 1 & & \\ CH_3 & — & CH_2 & — & CH_2 & — & CH_2 & — & Cl \end{array}$$

As expected, a primary hydrogen is less reactive than an ordinary secondary hydrogen, but among the three secondary carbons there is a striking decrease in reactivity on going from left to right. This reflects the inductive effect of the chlorine substituent, which destabilizes the transition state which has positively charged carbon.

Chlorination of propionic acid with Cl_2 (and a light initiator) also reveals the operation of a polar effect,[4] the hydrogens on a primary carbon being more reactive even though the secondary radical would also be conjugated with the carbonyl group. Both the per cent substitution on the two positions and the relative reactivity of a hydrogen on each carbon are indicated; the latter is corrected for the fact that there are three methyl hydrogens to only two methylene hydrogens.

$$\begin{array}{lcccl} \% & 70 & & 30 & \\ & CH_3 & — & CH_2 — CO_2H & \\ & 1 & : & 0.64 & \text{reactivity} \end{array}$$

Of course the polar effect is only one factor,[5] and it can be overcome in some cases. For instance,[4] chlorination of isobutyryl chloride gives 20% substitution next to the carbonyl, but when this result is corrected for the fact that there are six primary hydrogens compared with one tertiary one, it is seen that the tertiary hydrogen is really more reactive.

3. J. Hine, *Physical Organic Chemistry* (2nd ed., McGraw-Hill Book Co., New York, 1962), p. 459.
4. Ref. 2, p. 364.
5. For an interesting additional effect, cf. W. Thaler, "Photobromination of Alkyl Halides, an Unusual Orienting Effect in the Bromination of Alkyl Bromides," *Journal of the American Chemical Society,* **85,** 2607 (1963).

$$\begin{array}{cccc} \% & 80 & 20 & \\ & (CH_3)_2\text{—} & CH\text{—}COCl & \\ & 1: & 1.5 & \text{reactivity} \end{array}$$

In this case the greater stability of a tertiary conjugated radical is the dominant factor.

The importance of a polar effect should also depend on the electronegativity of the attacking radical. Methyl radicals can be generated by the decomposition of diacetyl peroxide; when this is done in isobutyryl chloride labeled with deuterium, it is found that the tertiary hydrogen is 12.4 times as reactive as any one of the primary hydrogens.[6]

$$CH_3\overset{O}{\overset{\|}{C}}O\text{—}O\overset{O}{\overset{\|}{C}}CH_3 \xrightarrow{\Delta} 2\ CH_3\overset{O}{\overset{\|}{C}}\text{—}O\cdot \longrightarrow CO_2 + CH_3\cdot$$

$$CH_3\cdot + (CH_3)_2CDCOCl \longrightarrow CH_4 + (\cdot CH_2)(CH_3)CDCOCl$$

$$\swarrow$$

$$CH_3D + (CH_3)_2C\cdot\text{—}COCl$$

Experimentally the relative reactivity is determined from the amount of deuterium which is found in the methane (after correction for a measured isotope effect). Polar effects play no role here since the attacking methyl radical has essentially the same electronegativity as the product radical.

One of the most interesting manifestations of polar effects in radical reactions is found in copolymerization.[7] When a radical polymerization chain is initiated in the presence of two different monomers, for instance acrylonitrile and acrylamide, it is fre-

6. C. Price and H. Morita, "The Reaction of Methyl Radicals with Isobutyryl and α-Deuteroisobutyryl Chloride," *Journal of the American Chemical Society,* **75,** 3686 (1953).
7. Ref. 2(a), Chapter 4.

quently found that a mixed polymer is formed, a so-called *copolymer,* in which the two monomers are randomly incorporated.

$$R\cdot + CH_2{=}CH{-}CN + CH_2{=}CH{-}CONH_2 \longrightarrow$$

$$-(CH_2CH(CN){-}CH_2{-}CH(CN){-}CH_2CH(CONH_2){-}){-}\text{etc.}$$

This is not surprising, for whenever a monomer has added to the chain a new radical end is formed which can now attack either monomer, and in this case their reactivities are quite similar so that they become incorporated more or less randomly. On the other hand, when two monomers of greatly different reactivity are used, such as the very reactive acrylonitrile and the much less reactive allyl alcohol, the reactive monomer will be incorporated first; the unreactive one will only seldom enter the polymer until the reactive monomer is used up.

$$R\cdot + CH_2{=}CH{-}CN + CH_2{=}CH{-}CH_2OH \longrightarrow$$

$$R{-}CH_2{-}CH(CN){-}CH_2{-}CH(CN){-}CH_2{-}CH(CN){-}CH_2CH(CN){-}\text{etc.}$$

These are the results expected without a consideration of polar effects, and they are the results which are often observed.

However, when styrene is copolymerized with maleic anhydride, it is found[7] that the reactivity of a monomer toward the growing

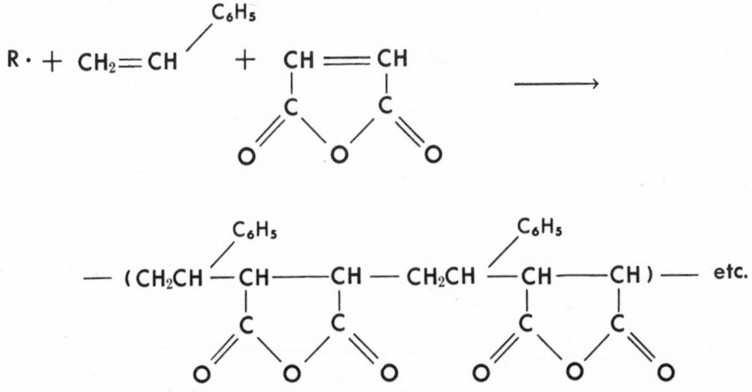

$$R\cdot + CH_2{=}CH(C_6H_5) + \underset{\text{(maleic anhydride: }O{=}C\text{—}O\text{—}C{=}O\text{ ring)}}{CH{=}CH} \longrightarrow$$

$$\text{—}(CH_2CH(C_6H_5)\text{—}CH\text{—}CH\text{—}CH_2CH(C_6H_5)\text{—}CH\text{—}CH)\text{— etc.}$$

(each CH—CH pair bonded to C(=O)—O—C(=O) forming an anhydride ring)

radical chain depends on the nature of the end of the chain. If a styrene unit has just been added, then there is a strong tendency for the chain to add to maleic anhydride, and when the anhydride has been added there is a strong tendency for the new end to attack styrene.

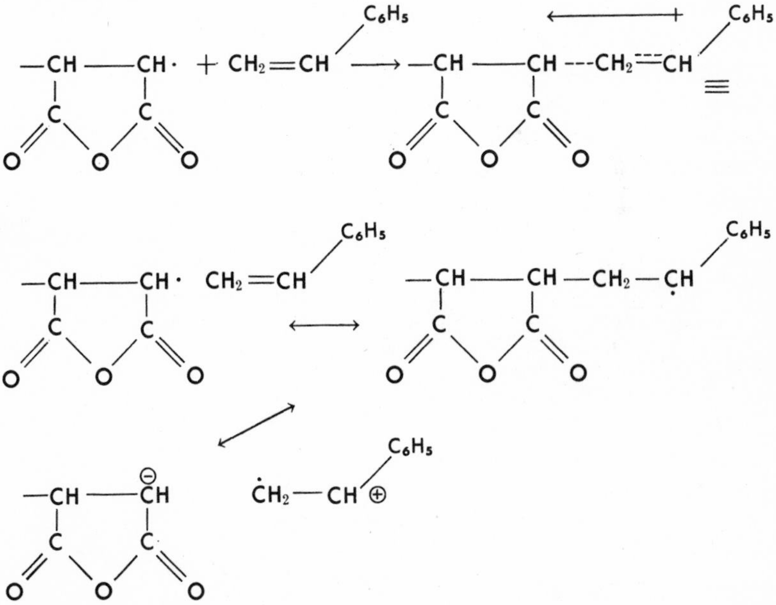

$$\text{—}CH\text{—}CH\cdot + CH_2{=}CH(C_6H_5) \longrightarrow \text{—}CH\text{—}CH\text{---}CH_2{\overset{---}{=}}CH(C_6H_5) \equiv$$

$$\text{—}CH\text{—}CH\cdot \quad CH_2{=}CH(C_6H_5) \longleftrightarrow \text{—}CH\text{—}CH\text{—}CH_2\text{—}\dot{C}H(C_6H_5)$$

$$\longleftrightarrow \quad \text{—}CH\text{—}\overset{\ominus}{C}H \quad \dot{C}H_2\text{—}CH(C_6H_5)^{\oplus}$$

(in each structure the CH—CH pair is part of the anhydride ring C(=O)—O—C(=O))

The result is that the polymer has more or less regular alternation of the two monomers.

The *alternating effect* is due to polar factors in the transition state for addition. When the end of the chain carries a maleyl radical, addition to styrene involves a transition state in which a polar form can make an appreciable contribution.

In the transition state the new bond is still quite long and the styryl double bond is not yet completely broken. Drift of electrons to the left occurs because the benzene ring can help stabilize a positive charge, and the carbonyl group is strongly electron-attracting. This extra ionic resonance form could not play a stabilizing role if maleic anhydride were being attacked instead, since maleic anhydride could not tolerate the partial positive charge. However, when a chain with a styryl end attacks maleic anhydride an ionic form contributes to stabilizing the transition state.

$$-CH_2-\overset{C_6H_5}{CH}\cdot \quad \underset{O=C-O-C=O}{CH=CH} \longleftrightarrow -CH_2-\overset{C_6H_5}{CH}-\underset{O=C-O-C=O}{CH-\dot{C}H}$$

$$\updownarrow$$

$$-CH_2-\underset{\oplus}{\overset{C_6H_5}{CH}} \quad \underset{O=C-O-C=O}{\dot{C}H-\overset{\ominus}{C}H}$$

It will be noted that this type of resonance form resembles the structures drawn in Special Topic 5 to explain the stability of charge-transfer complexes.

The alternating effect is quite commonly found when copolymerization involves two monomers with strongly different polar character. Such examples show that a simple treatment of reactivity in terms of bond energies and steric factors, as was done in Chapter 7, may have to be supplemented by consideration of the polar effects which can play a role in many radical reactions.

8

PHOTOCHEMISTRY

ALTHOUGH A SOLUTION of benzophenone in isopropyl alcohol is perfectly stable under ordinary conditions, irradiation with ultraviolet light causes the formation of acetone and benzpinacol.

$$2C_6H_5COC_6H_5 + CH_3\text{—}CHOH\text{—}CH_3 \xrightarrow{h\nu} \underset{\text{benzpinacol}}{(C_6H_5)_2\overset{\displaystyle OH}{\overset{|}{C}}\text{—}\overset{\displaystyle OH}{\overset{|}{C}}(C_6H_5)_2} + CH_3COCH_3 \quad (1)$$

In considering the mechanisms of such photochemical reactions we will be concerned with two questions:

1. What kind of new species is formed from the interactions of light with an organic molecule?
2. What is the sequence of subsequent changes which this species undergoes?

8–1 Excited States

As we pointed out in Chapter 7, ultraviolet light at 200 mμ has an energy of 143.2 kcal/mole, while even blue light at 400 mμ has 71.6 kcal/mole. These very large amounts of energy can break bonds and cause extensive changes in a molecule *if the light is absorbed.* In order that light be absorbed by a molecule there must be a quantum mechanically-permitted excited state of the molecule of the correct energy, and the absorption of light by exciting the molecule to this state must be "allowed." This latter requirement indicates that certain transitions between states are much less probable than others; we will see some of the most striking examples of this in considering the properties of triplet excited states. In addition to this direct mechanism, molecules may often be raised by *energy transfer* from another excited molecule.

With the kinds of energies available in ultraviolet light we are dealing with *electronic* excitation of molecules; the energies required to produce vibrational excitation are much smaller. In most cases we will be concerned with excited π electrons. For example, in ethylene absorption of light at 162 mμ leads to an excited state which has one of the π electrons promoted to the π antibonding state, π^*.

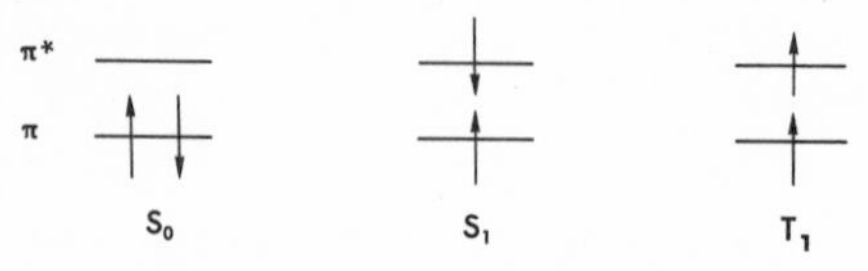

π—Electronic states of ethylene

The electron spins remain paired, so this excited state is a singlet, S_1. The ground state, which must have spins paired, is S_0. However, in an excited state with each orbital half occupied spins may also be unpaired, leading to the triplet state T_1. In more complex molecules with many occupied and unoccupied π orbitals there will be a number of possible higher energy singlet and triplet excited states.

Light of the correct energy can excite a molecule from its S_0 state to the S_1 or higher singlet state. In almost all cases, direct excitation of S_0 to a triplet state such as T_1 is quite improbable, and light of the correct wavelength for the $S_0 \rightarrow T_1$ change is only weakly absorbed. This is one of the selection rules alluded to earlier. Spinning electrons have angular momentum, and the change from S_0 to T_1 would involve a change in angular momentum as one of the spins is reversed. Conservation of total angular momentum is a fundamental law of nature, and during the 10^{-15} seconds required for the electronic transition it is quite improbable that some other interaction can occur so as to compensate for the change in spin angular momentum on going from S_0 to T_1 and allow the total momentum to stay constant. Light can only be absorbed for an $S_0 \rightarrow T_1$ transition when such an improbable compensating interaction occurs.

Of course all the electronic states of a molecule have vibrationally-excited levels associated with them. Excitation could occur from S_0 to various vibrational states of S_1, so absorption spectra generally consist not of a single line but of a series of lines corresponding to energy differences between each of the populated vibrational states of S_0 and the various vibrational levels of S_1. In most spectra this series of lines coalesces into a broad absorption band. According to the *Franck–Condon principle* atoms don't have time to move during the 10^{-15} seconds involved in absorption of light, so the excited molecule initially has the same geometry as in the ground state; but in 10^{-12} seconds or so after absorption of light the molecules in upper vibrational states of S_1 cascade down to the lowest levels of S_1 and adopt the new equilibrium geometry of S_1 by dissipating their small amounts of excess vibrational energy to the environment. Even molecules in higher electronic singlet states (S_2, etc.) usually cascade rapidly down to the lowest levels of S_1. Accordingly, we can often refer to "the excited singlet state"—meaning S_1 in vibrational equilibrium with its surroundings—even though excitation at first may have been to a higher energy singlet.

One could imagine a further cascade of S_1 down to S_0, passing successively down the vibrational levels of S_0 to the ground state with dissipation of the energy as heat. This process, an example of *internal conversion* (radiationless conversion between different electronic states with the same total spin), happens to some degree for all molecules. However, the energy difference between S_1 and S_0 is often quite large, and the probability of transferring large amounts of energy to the environment in quantum mechanically allowed ways is limited, so S_1 has an appreciable lifetime.

S_1 may emit the energy as a photon and drop to (some vibrational level of) S_0. This process, *fluorescence,* generally occurs within 10^{-9} to 10^{-6} seconds. During this time the excited molecule may instead undergo internal conversion, as discussed above, or it may undergo chemical reactions. The latter must be quite rapid since the lifetime of S_1 before fluorescence is so short, and even diffusion-controlled bimolecular reactions would require 10^{-10} seconds or so. Of course, unimolecular chemical reactions of S_1 or facile bimolecular reactions with other materials in high concentration, such as solvent, are possible in the time available.

The other process of importance for S_1 is *intersystem crossing,* conversion of a singlet to a triplet state (or vice versa). In some molecules S_1 can convert to T_1 before fluorescence occurs. The difference in spin angular momentum is not a prohibitive problem since during the lifetime of S_1 there is a finite opportunity to compensate for it with changes in angular momentum elsewhere in the system; energy must also be dissipated since triplet states are almost always lower in energy than the corresponding singlets, but S_1 may first be converted to a vibrationally-excited T_1 of similar energy with subsequent cascade to the lowest levels of T_1. The efficiency of intersystem crossing varies with molecular structure in a way which is not completely understood. Benzophenone, the molecule we will be considering in detail, undergoes the process with almost unit efficiency, so the excited singlets are all converted to the triplet. The possible processes we have discussed are illustrated in Fig. 8–1.

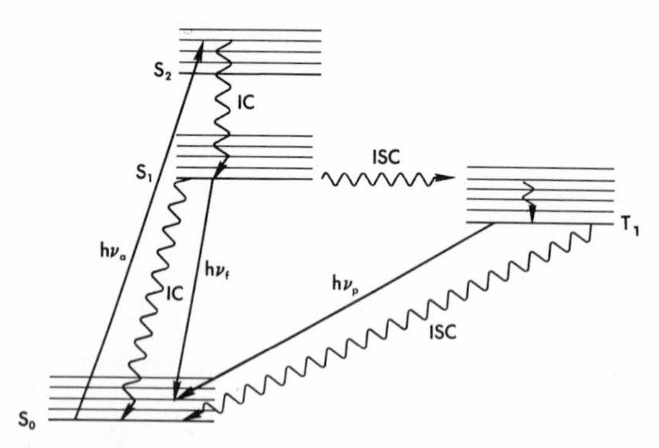

FIGURE 8–1

Excitation, fluorescence, phosphorescence, and radiationless transitions. Light of energy $h\nu_a$ is absorbed promoting the molecule from its ground state S_0 to an excited singlet S_2. Internal conversion (IC) of this to the lowest vibrational state of S_1 may be followed by fluorescence in which $S_1 \rightarrow S_0 + h\nu_f$, internal conversion of S_1 to S_0 without radiation, or intersystem crossing (ISC) of S_1 to T_1. The latter may phosphoresce (losing energy $h\nu_p$) or undergo a radiationless intersystem crossing to S_0. Wavy lines indicate radiationless processes while solid arrows show transitions involving absorption or emission of photons.

The triplet state T_1 may not convert to S_1 again since T_1 is of lower energy. It can go to S_0 by emission of a photon, but this process is quite slow because of the spin angular momentum problem, so triplet lifetimes before such *phosphorescence* may range from 10^{-3} seconds to 10 seconds or even longer. The triplet state may undergo intersystem crossing to S_0, the energy being dissipated as heat, but this is again slower than the corresponding $S_1 \rightarrow S_0$ process because of the change in spin. The triplet state T_1 may also enter chemical reactions; the time available is much longer than for S_1, so photochemistry involving triplet states is not restricted to very fast processes and is more extensive than for singlets.

8–2 Energy Transfer

There is one other process available to S_1 and T_1 to which we have not referred. These excited species may transfer their energy to molecules in the environment not just in small bits but all at once. The acceptor molecule is thereby raised to an electronically excited state of its own. This can be detected first of all in the *quenching* of the fluorescence or phosphorescence of molecule A by the addition of acceptor B. Secondly, the excited molecule B may fluoresce or phosphoresce itself, or undergo chemical changes. Considered from the standpoint of B, A is a *photosensitizer;* it absorbs the light and transfers the energy to B.

This is particularly important in the case of *triplet* energy transfer. The triplet state of A may transfer both

$$A(T_1) + B(S_0) \rightarrow A(S_0) + B(T_1)$$

energy and spin angular momentum to B, so total angular momentum is conserved. Such transfer occurs at a diffusion-controlled rate provided it is energetically downhill (the $S_0 \rightarrow T_1$ difference is greater for A than for B). Since T_1 is not available for many molecules by direct processes ($S_0 + h\nu \rightarrow T_1$ is "forbidden," and $S_1 \rightarrow T_1$ does not always occur), photosensitization is quite important in photochemistry.

8–3 Photochemical Reductions

Let us now return to a consideration of Reaction (1), the photochemical reduction-dimerization of benzophenone by isopropyl alcohol. The solution is irradiated with ultraviolet light with a wavelength distribution from 300–350 mμ. In this region isopropyl alcohol does not absorb light, but benzophenone undergoes an $n \rightarrow \pi^*$ singlet transition with λ_{max} near 350 mμ. The π^* orbital is the antibonding π orbital of the carbonyl group, although it is conjugated with the phenyl groups and more stable than π^* of aliphatic ketones. The electron which is promoted to the π^* orbital is not one of the π electrons of benzophenone, but is one of the so-called n electrons of the unshared pairs on the carbonyl oxygen. The π electrons are of lower energy than the unshared pairs, so $\pi \rightarrow \pi^*$ excitation in benzophenone requires more energy and occurs at 250 mμ. The energy of (delocalized) π electrons depends on structure, and there are some other ketones for which the $\pi \rightarrow \pi^*$ excitation rather than $n \rightarrow \pi^*$ leads to the lowest singlet state S_1.

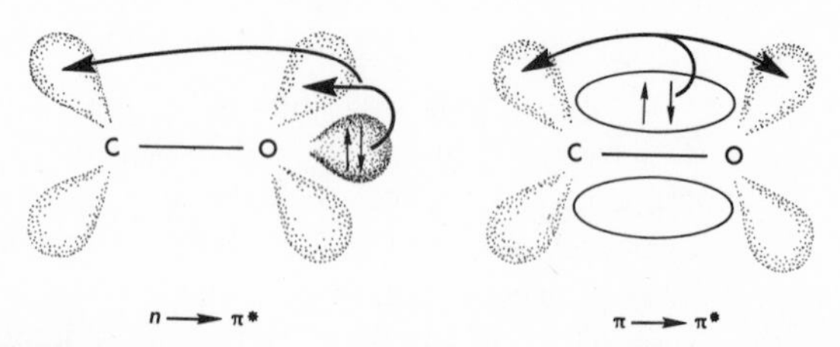

In benzophenone the state with one of the oxygen n electrons in a π^* orbital, S_1, undergoes very rapid intersystem crossing to T_1 (the lifetime of S_1 is ca 10^{-10} seconds), and the triplet is actually the species which reacts with isopropyl alcohol. One piece of evidence that it is T_1, not S_1, which attacks the isopropyl alcohol is the effect of quenchers on the *quantum yield.*

The quantum yield of a reaction, Φ, is the ratio of product molecules produced to photons absorbed. If each photon absorbed leads to the formation of a product molecule, then Φ will be 1.0, while dissipation of some of the light energy by fluorescence, phosphorescence, or radiationless cascades to S_0 will reduce the quantum yield. In the case of chain reactions, such as the free radical chlorination initiated by light which was discussed in Chapter 7, quantum yields can be much larger than 1. The quantum yield of Reaction (1) varies with the concentration of benzophenone, and at high concentrations it is close to 1.0, so this process can be quite efficient. However, the addition of a small concentration of naphthalene to the solution results in a drastic drop in the quantum yield, although its spectrum and that of the benzophenone show that the naphthalene is not absorbing any of the light. Naphthalene is instead acting as a quencher of an excited state of benzophenone, the energy transfer causing a return of benzophenone to S_0 with simultaneous excitation of the naphthalene. The naphthalene must be quenching a triplet state, not a singlet, since only the T_1 state of benzophenone would have a lifetime long enough to undergo collisions with very low concentrations of naphthalene. Thus the singlet states must be going to T_1 before chemical reaction occurs. By the technique of *flash photolysis* benzophenone triplet has also been directly observed to be the species which attacks isopropyl alcohol.

$$(C_6H_5)_2\overset{\uparrow}{C}{=}\ddot{O}\uparrow \longleftrightarrow (C_6H_5)_2\overset{\uparrow}{C}{-}\ddot{\underset{..}{O}}\uparrow$$

Two resonance forms of benzophenone $n \rightarrow \pi^*$ triplet, T_1

The $n \rightarrow \pi^*$ triplet has a remaining odd electron in an oxygen n orbital which makes it resemble an oxygen free radical. Like other oxygen radicals it can remove the α-hydrogen of isopropyl alcohol leading to two stabilized radicals.

$$\underset{T_1}{(C_6H_5)_2\overset{\uparrow}{C}{-}\ddot{\underset{..}{O}}\uparrow} + (CH_3)_2CHOH \longrightarrow \underset{\text{I}}{(C_6H_5)_2\dot{C}{-}OH} + \underset{\text{II}}{(CH_3)_2\dot{C}{-}OH}$$

The radical I dimerizes to benzpinacol, while II can disproportionate to acetone and isopropyl alcohol. However, this mechanism would lead to a maximum quantum yield Φ for benzpinacol formation of only 0.5, since two such reactions are needed to form the two molecules of I for one dimer. At high concentrations of benzophenone, Φ goes to 1.0, so some other process must also be involved. Apparently II can react with benzophenone to produce another molecule of I, so that two molecules of I are produced per photon absorbed.

$$\underset{\text{II}}{(CH_3)_2\dot{C}OH} + (C_6H_5)_2C{=}O \rightarrow (CH_3)_2C{=}O + \underset{\text{I}}{(C_6H_5)_2\dot{C}{-}OH}$$

The reaction occurs because I is more resonance stabilized than II.

Although few reaction mechanisms have been examined in as much detail as that of Reaction (1), photoreductions are known with a number of other ketones and hydrogen donor combinations.

8-4 Photochemical Cycloadditions

When benzophenone is irradiated in solution with isobutylene, the benzophenone $n \rightarrow \pi^*$ triplet is again formed, and to some extent it removes allylic hydrogen atoms from the olefin to give photoreduction. However, the major product is an oxetane, from two-step cycloaddition of the triplet to the olefin. Since the intermediate III has an unpaired electron on a tertiary carbon,

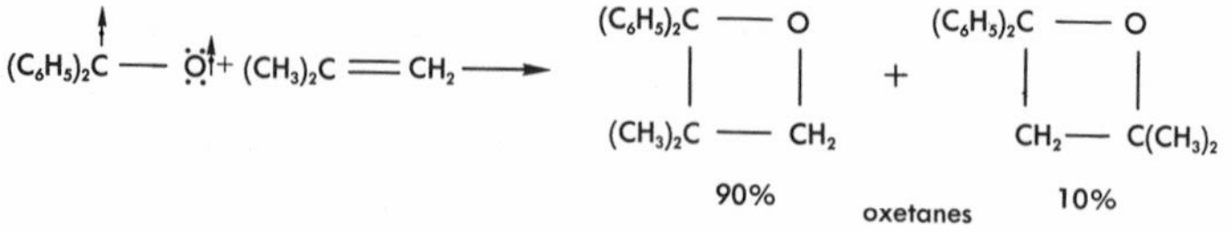

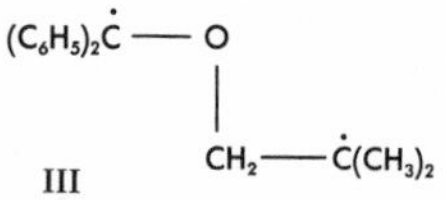

$$(C_6H_5)_2\overset{\uparrow}{C}-\ddot{\overset{\uparrow}{O}}: + (CH_3)_2C{=}CH_2 \longrightarrow \begin{matrix}(C_6H_5)_2C & - & O \\ | & & | \\ (CH_3)_2C & - & CH_2\end{matrix} + \begin{matrix}(C_6H_5)_2C & - & O \\ | & & | \\ CH_2 & - & C(CH_3)_2\end{matrix}$$

90% oxetanes 10%

$$\begin{matrix}(C_6H_5)_2\dot{C} & - & O \\ & & | \\ & & CH_2 - \dot{C}(CH_3)_2\end{matrix}$$

III

there is some preference (90%) for cycloaddition to occur in the direction shown, but the isomeric oxetane is formed in 10% yield. Intermediate III when first formed is still a triplet, and formation of the second bond requires spin inversion.

One might expect that cycloaddition of benzophenone triplet to a conjugated diene would be even easier than to a simple olefin, since the intermediate radical would have allylic resonance stabilization. However, irradiation of benzophenone with butadiene (again at the wavelengths used the light is absorbed only by the benzophenone) leads not to oxetane formation but to dimerization of the butadiene.

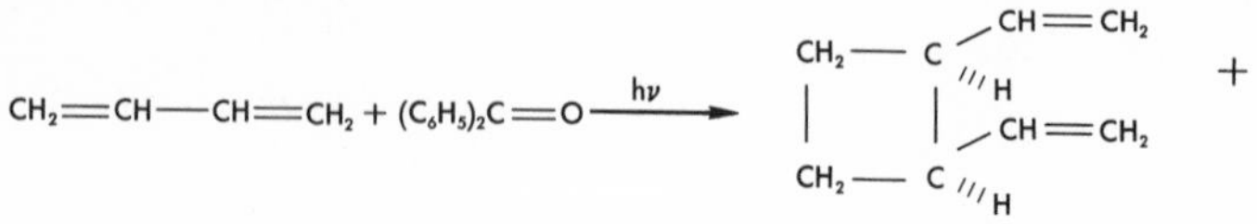

$$CH_2{=}CH-CH{=}CH_2 + (C_6H_5)_2C{=}O \xrightarrow{h\nu} \begin{matrix}CH_2 & - & C(H)(CH{=}CH_2) \\ | & & | \\ CH_2 & - & C(H)(CH{=}CH_2)\end{matrix} +$$

IV

$$\begin{matrix}CH_2 & - & C(H)(CH{=}CH_2) \\ | & & | \\ CH_2 & - & C(H)(CH{=}CH_2)\end{matrix} + \text{cyclohexene ring: } CH_2-CH{=}CH-CH_2-CH_2-CH(-CH{=}CH_2)$$

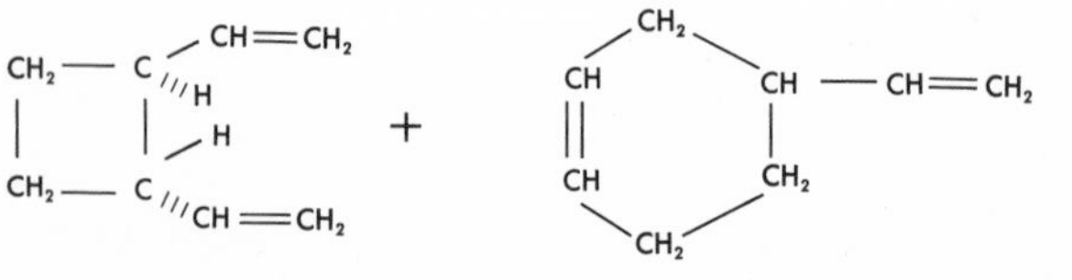

V VI

When benzophenone triplet collides with butadiene, a triplet energy transfer occurs, so the ketone is acting as a photosensitizer. The triplet state of butadiene is of lower energy than that of benzophenone (relative to their ground states), so triplet energy transfer is possible here but not for a simple olefin (which would require a higher energy sensitizer); when energy transfer is energetically allowed, it is always faster than chemical bond formation. Butadiene triplet cannot be formed without benzophenone since $S_0 + h_\nu \rightarrow T_1$ is strongly forbidden. The S_1 state of butadiene can be formed with light of higher energy than 300 mμ, but even if such high energy light is used, intersystem crossing is quite inefficient in the diene, and the excited singlets would not tend to form much triplet. In a real sense the triplet state of butadiene is available for reaction only through the use of triplet photosensitizers such as benzophenone.

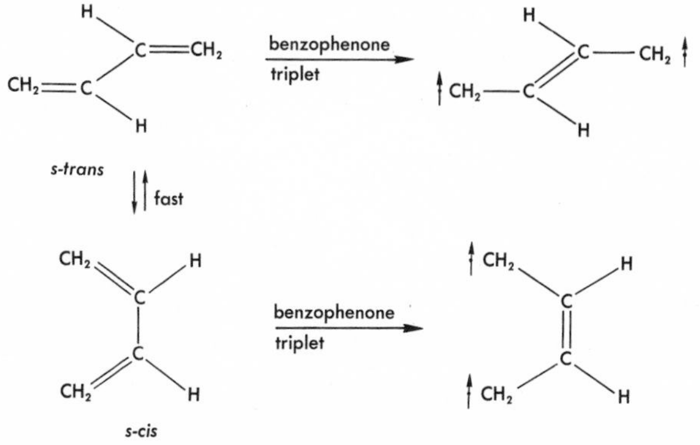

The triplet butadiene now adds to ordinary butadiene by two-step mechanisms to give the three dimers. However, the remarkable observation has been made that the relative amounts of IV, V, and VI formed depends on the exact photosensitizer used. It is straightforward to determine the triplet energies of various photosensitizers (by measuring the wavelength of their phosphorescence), and it turns out that sensitizers with a triplet energy greater than 60 kcal/mole—including benzophenone with E_T of 69 kcal/mole—produce only a few per cent of the Diels–Alder type product VI, with the remainder a 1:4 ratio of IV and V. As E_T of the photosensitizer is decreased to 53 kcal/mole, the 1:4 ratio of IV and V stays constant, but over 40% of the product is now VI. Below E_T of 53 kcal/mole the quantum yield falls off rapidly since excitation of the diene now requires more energy than the sensitizer has.

Butadiene exists in the *s-trans* conformation in solution, with a few per cent of *s-cis* as well. When butadiene accepts triplet energy and undergoes a $\pi \rightarrow \pi^*$ excitation to T_1, it forms a triplet with defined stereochemistry. Although the particular resonance forms we have written indicate that rotation between carbons 2 and 3 is restricted in the triplets, a consideration of the molecular orbitals of butadiene (Figure 8–2) will also make this clear.

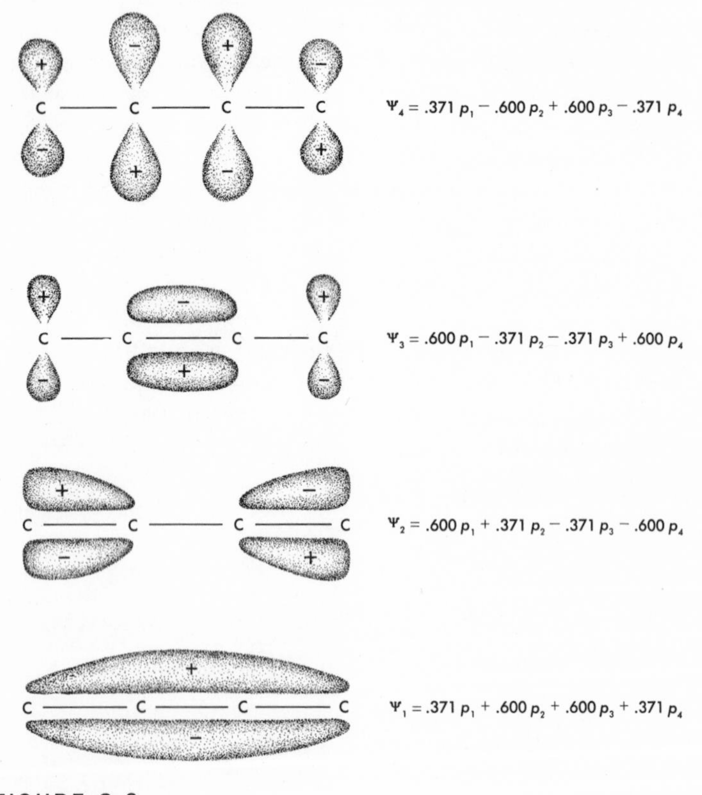

$\Psi_4 = .371\,p_1 - .600\,p_2 + .600\,p_3 - .371\,p_4$
$\Psi_3 = .600\,p_1 - .371\,p_2 - .371\,p_3 + .600\,p_4$
$\Psi_2 = .600\,p_1 + .371\,p_2 - .371\,p_3 - .600\,p_4$
$\Psi_1 = .371\,p_1 + .600\,p_2 + .600\,p_3 + .371\,p_4$

FIGURE 8–2

Molecular π orbitals of butadiene.

The lowest $\pi \rightarrow \pi^*$ transition involves removing an electron from Ψ_2, which is antibonding between carbons 2 and 3, and placing it into Ψ_3, which has a 2—3 bond. The result (together with the 2—3 bonding of the electrons in Ψ_1) is a strong 2—3 π bond which prevents rotation during the lifetime of the butadiene triplet.

The energy change involved in going to the triplet state can be different for the *cis* and *trans* conformations of butadiene, and other evidence indicates that E_T is 60 kcal/mole for *trans* and 53 kcal/mole for *cis*. A high-energy sensitizer can transfer energy to either conformation; the *trans* triplet is thus formed predominantly, because the diene is mostly *trans*. A sensitizer with E_T below 60 kcal/mole but above 53 kcal/mole will preferentially form the *cis*-butadiene triplet, since energy transfer will only occur readily to *s-cis*-butadiene.

Attack of a *trans*-butadiene triplet on *s-trans*-butadiene leads to an intermediate hexadienyl diradical which still does not have free rotation about the erstwhile 2—3 bonds, so it cannot close to VI. The *cis*-butadiene triplet can give VI. Of course, in both cases a spin must invert before the second bond can form.

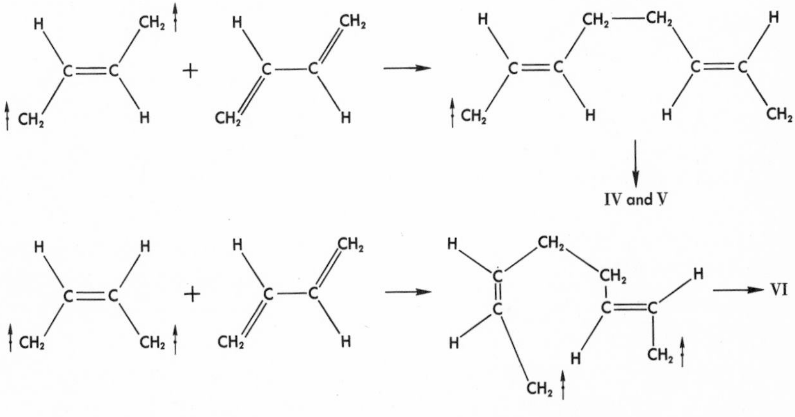

IV and V

VI

Generally triplet state photosensitizers are used in this way to produce other *triplets,* but there is one important exception. The O_2 molecule exists in its ground state as a triplet, and its lowest excited state is a singlet. Direct conversion of O_2 to this excited singlet state with light is again a forbidden process, but singlet oxygen can be produced using a triplet photosensitizer since total spin is thus conserved during energy transfer. This singlet excited-state O_2 can add in one step to dienes, as in a Diels–Alder reaction. Singlet O_2 has also been generated by chemical means, and it

$$\uparrow\ddot{\underset{..}{O}}-\ddot{\underset{..}{O}}\uparrow + \downarrow A^{*}\downarrow \longrightarrow \left[\uparrow\ddot{\underset{..}{O}}-\ddot{\underset{..}{O}}\downarrow\right]^{*} + A\downarrow\uparrow$$

T_0 T'_1 S_1 S'_0

behaves identically with that generated with light and sensitizer.

$$\text{(cyclohexadiene)} + O_2 \xrightarrow[h\nu]{\text{Sensitizer}} \text{(endoperoxide)}$$

8–5 Photoisomerizations

If a dilute solution of *trans*-1,3-pentadiene is irradiated in the presence of benzophenone, it is converted to a mixture of 55% *trans* and 45% *cis* pentadiene (dimerization is of course suppressed by dilution). This same mixture is also formed starting with the pure *cis* diene. The two dienes can be interconverted through the

$$\text{trans-1,3-pentadiene} \underset{\text{benzophenone}}{\overset{h\nu}{\rightleftharpoons}} \text{cis-1,3-pentadiene}$$

trans - 1, 3 - pentadiene *cis* - 1, 3 - pentadiene

triplet of the diene, since in this triplet the 1—2 (and 3—4) π bonds are weakened. This is seen by considering the molecular orbitals of butadiene in Figure 8–2; an electron is removed from Ψ_2 (1—2 bonding) and placed in Ψ_3 (1—2 antibonding). The *photostationary state* comes when the rate of excitation of the *cis* diene equals its rate of formation from excited states. The composition of the photostationary mixture thus depends on a number of rate constants, including those for energy transfer from the photosensitizer to the dienes. For *cis* pentadiene E_T is lower than for *trans,* so with low energy sensitizers the *cis* diene is selectively excited, and the photostationary composition has a higher proportion of *trans* diene than the 55% with benzophenone.

Direct irradiation of *trans, cis, trans*-2,4,6-octatriene leads to a different sort of photostationary equilibrium, with its valence isomer *trans*-5,6-dimethylcyclohexadiene. This reaction is completely stereospecific in both directions as shown, for very interesting reasons to be discussed in Special Topic 8.

A number of more complex photochemical rearrangements have been discovered and studied, and more are being found each day. However, an extensive consideration of the field is beyond the scope of this book, and this chapter has been devoted instead to illustrating fundamental principles which are common to all photochemical processes.

General References

N. J. Turro, *Molecular Photochemistry* (W. A. Benjamin, New York, 1965). The best introduction to this topic.

J. G. Calvert and J. N. Pitts, *Photochemistry* (John Wiley & Sons, New York, 1966). A more advanced, and encyclopedic, treatment.

R. O. Kan, *Organic Photochemistry* (McGraw-Hill Book Co., New York, 1966). Surveys a variety of photochemical reactions at an introductory level.

D. C. Neckers, *Mechanistic Organic Photochemistry* (Reinhold, New York, 1967). Similar to the books by Turro and Kan.

W. A. Noyes, G. S. Hammond, and J. N. Pitts, *Advances in Photochemistry* (John Wiley & Sons, New York). A series, started in 1963, in which recent advances are discussed.

8

Special Topic

▪ ORBITAL SYMMETRY RELATIONSHIPS IN THERMAL AND PHOTOCHEMICAL REARRANGEMENTS[1]

ON HEATING, cyclobutene derivatives undergo a stereospecific rearrangement to butadiene derivatives. Thus *cis*-3,4-dimethylcyclobutene affords only *cis, trans*-1,4-dimethylbutadiene while the *trans*

1. A simple introduction to some of the ideas involved in this treatment is presented by J. Vollmer and K. Servis, "Woodward–Hoffmann Rules: Electrocyclic Reactions," *Journal of Chemical Education,* **45,** 214 (1968) and by M. Orchin and H. Jaffé, *The Importance of Antibonding Orbitals* (Houghton-Mifflin, Boston, Massachusetts, 1967). A more difficult treatment is that by R. Hoffmann and R. B. Woodward, "The Conservation of Orbital Symmetry," *Accounts of Chemical Research,* **1,** 17 (1968).

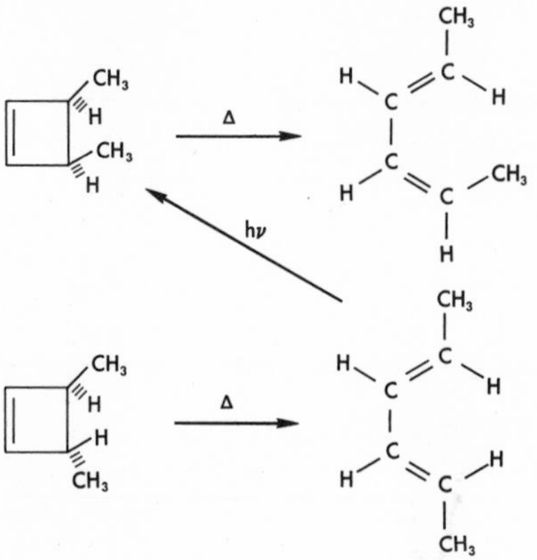

dimethylcyclobutene goes to *trans, trans*-dimethylbutadiene.[2] By contrast, on irradiation *trans, trans*-dimethylbutadiene closes to *cis*-3,4-dimethylcyclobutene.[3] In Chapter 8 we referred to a related case, the stereospecific photochemical interconversions of a cyclohexadiene and a hexatriene. In this special topic we will consider the reasons for such stereospecificity. Thermal reactions of various kinds will be considered first.

Cyclobutenes

Let us consider an orbital symmetry correlation diagram of the type introduced in Special Topic 4. Thus both cyclobutene and *s-cis*-butadiene (the conformation which is formed first) have a plane of symmetry perpendicular to the molecular plane. If in a *cis*-disubstituted cyclobutene the groups on the two saturated carbons rotate in opposite directions (the right one clockwise and the left one counterclockwise), then the plane of symmetry is maintained throughout the conversion to butadiene. For such a transition state to be stable it should be intermediate between

2. R. E. Winter, "The Preparation and Isomerization of *cis*- and *trans*-3,4-Dimethylcyclobutene," *Tetrahedron Letters,* 1207 (1965).
3. R. Srinivasan, "Mechanism of the Photochemical Valence Tautomerization of 1,3-Butadienes," *Journal of the American Chemical Society,* **90,** 4498 (1968).

the electronic ground states of starting material and product; as we pointed out previously, this will not be possible if the starting m.o.'s and product m.o.'s have different symmetries. Relative to this plane the σ single bond of cyclobutene is S and the π bond as well. These two orbitals cannot smoothly convert to the ground state orbitals of butadiene, however, by a *disrotatory* process (rotation of the saturated carbons in opposite directions) which maintains the symmetry plane: Relative to this plane Ψ_1 of butadiene is S, but Ψ_2 is A (cf. Figure 8–2 for the butadiene orbitals). The excited-state orbital Ψ_3 is S with respect to the plane between carbons 2 and 3; thus in the transition state for disrotatory ring opening one of the electron pairs can be in an S orbital which resembles both σ and Ψ_1, but the other pair is in an S orbital which resembles π and Ψ_3 (Figure 8–3). The orbital Ψ_3 is of high energy. Alternatively, the second electron pair could go into an A transition-state orbital, so it resembled the more stable Ψ_2, but then it would also have some of the character of π^* in cyclobutene and thus again be of high energy. It is because of these problems that Reactions (1) and (2) occur by a *conrotatory* path.

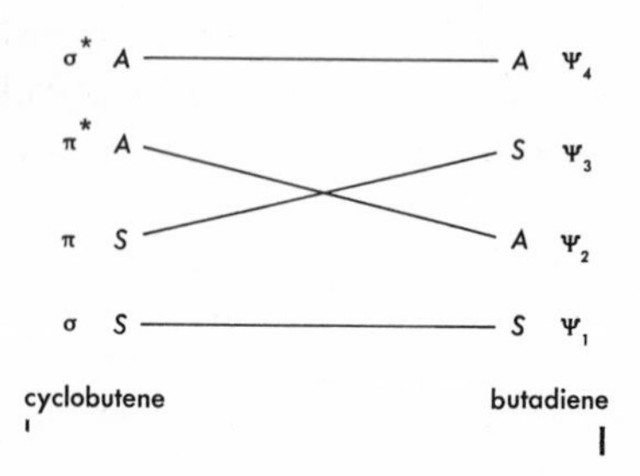

FIGURE 8–3
Orbital symmetry correlations for disrotatory interconversions of cyclobutene and butadiene.

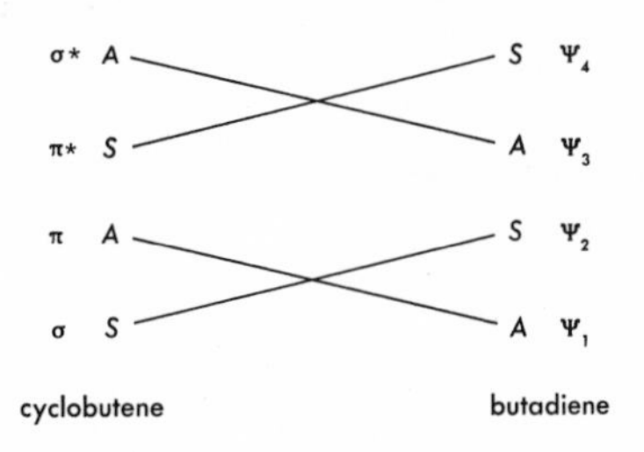

FIGURE 8–4
Orbital symmetry correlations for conrotatory interconversion of cyclobutene and butadiene.

The stereochemical results indicated in Reactions (1) and (2) show that the saturated cyclobutene carbons rotate in the same direction, so the change is called *conrotatory* instead of *disrotatory*.

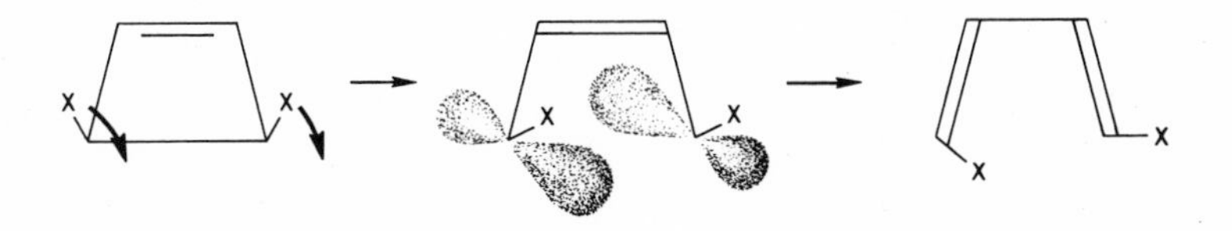

The transition state for a conrotatory change does not have the plane of symmetry we have been considering. Consequently its wave functions are neither *S* nor *A* with respect to such a plane, and it is possible for them to resemble to some extent the ground-state wave functions of both starting material and product.

Although the transition state for conrotatory ring opening does not have the plane of symmetry common to cyclobutene and buta-

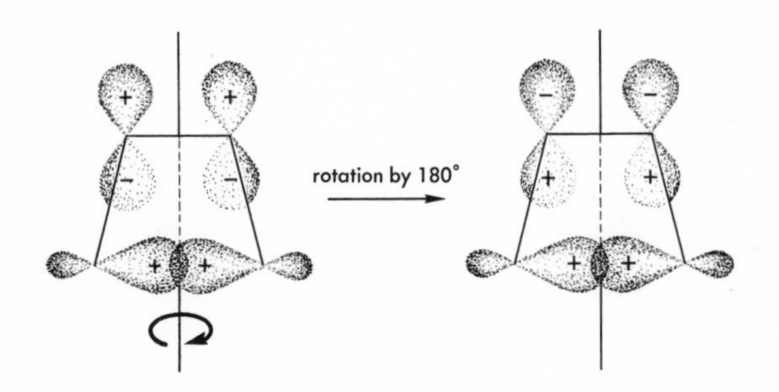

diene, it does share another symmetry element with them. All three have a two-fold rotational axis of symmetry: Rotation of cyclobutene by 180° around a line passing through the centers of the 1—2 and 3—4 bonds leads to an identical atomic arrangement.

Under this symmetry operation the σ bond is *S*, but the π bond has the signs reversed, so it is *A*. On the other hand, σ^* is *A* both under this rotation and under the previously discussed reflection through a plane. The symmetries of the cyclobutene and butadiene orbitals under this rotation are shown in Figure 8–4. With respect to this rotational axis both ground states have an *S* and an *A* orbital. The transition state in a conrotatory change also can have one *S* and one *A* orbital with respect to this axis, so it can correlate with the ground states on both extremes and need not be of high energy.

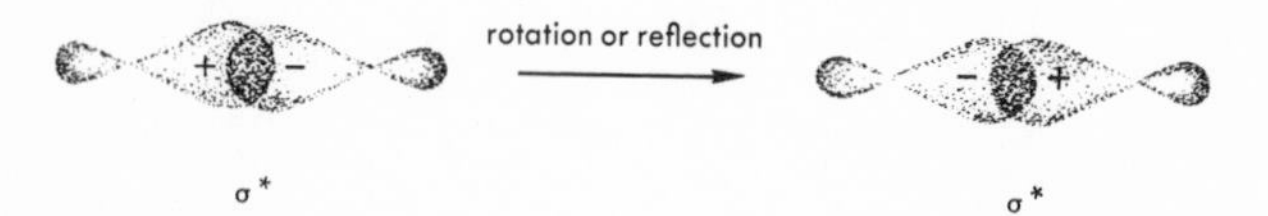

Such symmetry correlation arguments are not always satisfying, and they are not strictly applicable when unsymmetrical molecules are involved. In particular, the reader will have noticed that while cyclobutene may have both the plane and the axis of symmetry, *cis*-3,4-dimethylcyclobutene has only the plane, while *trans*-3,4-dimethylcyclobutene has only the rotational axis. The perturbation of molecular symmetry when a hydrogen is replaced by methyl is not likely to cause drastic changes in chemistry, and in fact Reactions (1) and (2) both involved conrotatory changes, so some more general argument than strict symmetry correlations must be invoked.

The interconversion of butadiene and cyclobutene is an example of an *electrocyclic* process. In such a process interconversion occurs between one isomer with n π electrons and the other with $n - 2$ π electrons and two electrons in a new σ bond. In the course of the change each product orbital is being formed by transformation of a particular starting material orbital; the transformation occurs by a smooth continuous increase of electron density at some atoms and decrease at others. The new π orbitals in the cyclic isomer are derived from π orbitals of the open-chain isomer in a straightforward way: The lowest energy π orbital of starting material and product interconvert (each is nodeless), the next highest π orbitals interconvert (each has one node in the center of the π system), etc. Thus the π orbitals of the cyclic isomer are derived from the lowest energy π orbitals of the open-chain isomer, and *the σ orbital of the cyclic isomer is derived from the highest π orbital of the open-chain isomer.*

As we have seen in Figure 8–2, the π orbital of cyclobutene is related to Ψ_1 of butadiene, since only Ψ_1 is bonding between carbons 2 and 3. The orbital Ψ_2 has an antibond where the π bond of cyclobutene is to develop, so it cannot contribute to π, but with conrotation Ψ_2 can lead to the σ bond. Disrotation is not allowed since Ψ_2 would then form a σ^* antibond. Even with unsymmetrical derivatives, Ψ_1 and Ψ_2 preserve their essential character of being bonding and antibonding respectively between carbons 2 and 3, so the argument is general.

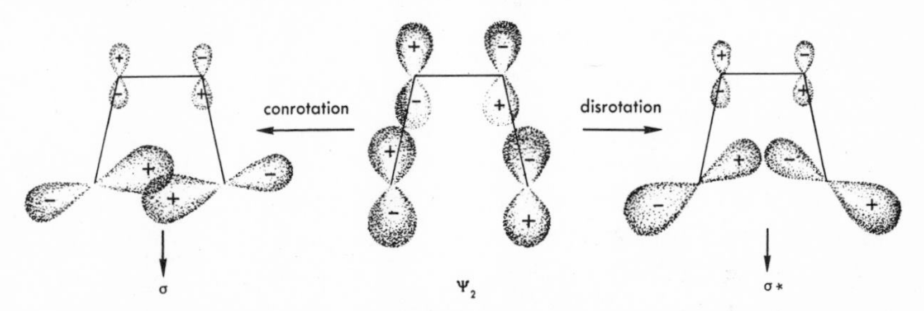

Finally, it might be noted that disrotation leads to the kind of upper lobe-upper lobe overlap which is similar to normal π bonding; as we saw in Special Topics 1 and 4, this kind of bonding is unfavorable when 4 electrons are cyclically delocalized, but it is favorable when 2, 6, and other $4n + 2$ numbers of electrons are involved. Accordingly, we might expect the stereochemical situation to change in a six-electron case.

Cyclohexadienes

When *trans, cis, trans*-1,6-dimethylhexa-1,3,5-triene is heated, it cyclizes to *cis*-1,2-dimethylcyclohexa-3,5-diene (reaction 3).[4]

4. E. N. Marvell, G. Caple, and B. Schatz, "Thermal Valence Isomerizations . . . ," *Tetrahedron Letters,* 385 (1965). This reaction is also reported by E. Vogel, W. Grimme, and E. Dinné, *ibid.,* 391 (1965).

(3)

This is the result of a disrotatory closure, the opposite of the cyclobutene ⇄ butadiene pathway. This stereochemistry can be understood by simply noting that with six electrons to be cyclically delocalized, normal overlap between the 1 and 6 carbons of a hexatriene will be stabilizing, and such overlap occurs with disrotation. Alternatively, we may consider the symmetries, bonds, and antibonds of the three occupied hexatriene molecular π orbitals. As

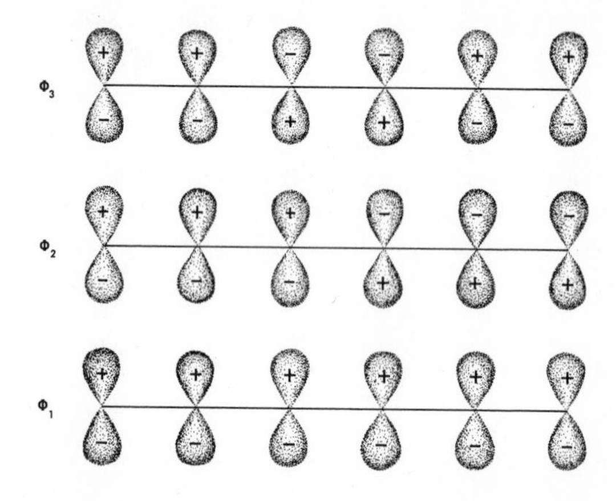

is the general pattern in polyenes the lowest orbital has no antibonds, the second has one antibond in the middle, and the third has two symmetrically-placed antibonds. Carbons 1 and 6 form a σ bond on cyclization, while carbons 2 through 5 form the diene system in the cyclic isomer. This diene system will have a Ψ_1 which

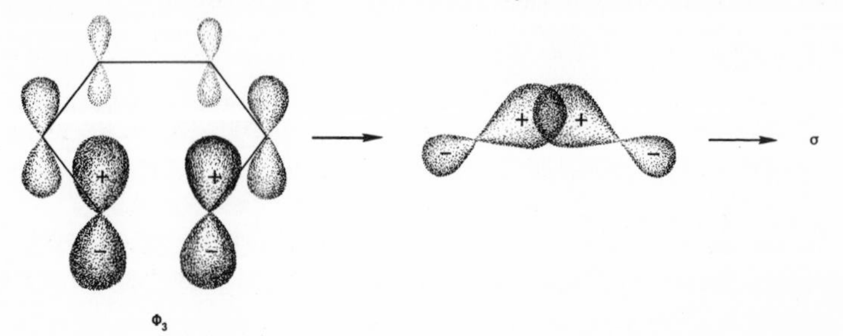

is all bonding and a Ψ_2 with one antibond in the middle, as in butadiene, so $\Phi_1 \rightarrow \Psi_1$ and $\Phi_2 \rightarrow \Psi_2$. Again it is seen that the new σ bond must be derived from the highest occupied π orbital of the starting triene, Φ_3, and disrotation is needed to form this σ bond. Finally, symmetry correlation diagrams like those in Figures 8–3 and 8–4 can be constructed which also show that in this case disrotation is favored. This is left to the reader.

It should be mentioned in passing that the electrocyclization of hexatriene derivatives is intellectually related to the Cope rearrangement of 1,5-hexadienes.[5]

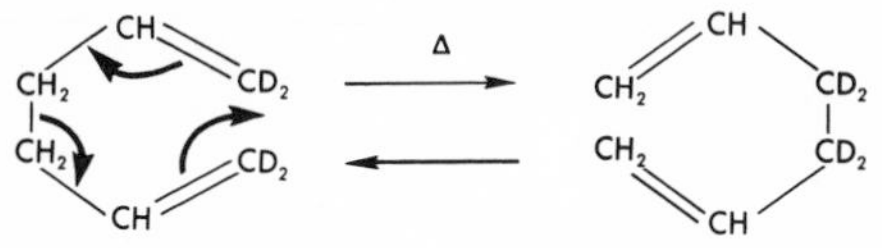

In this case a σ, not a π, bond is broken in forming the new bond, but again 6π electrons are delocalized in the transition state. Such processes can be very rapid if all the atoms are held in the right position by other bonds.

Cyclopropyl Cation

Cyclopropyl cation undergoes a ring opening to allyl cation; again orbital symmetry considerations predict a stereospecific process, disrotation.[6]

The argument is quite straightforward since only two electrons are involved, in the σ orbital of cyclopropyl cation and the completely bonding lowest π orbital (Special Topic 1) of allyl cation.

5. S. J. Rhoads, "Rearrangements Proceeding through 'No Mechanism' Pathways," in *Molecular Rearrangements,* P. de Mayo, ed. (John Wiley & Sons, New York, 1963), Vol. I, p. 655.
6. R. B. Woodward and R. Hoffmann, "Stereochemistry of Electrocyclic Reactions," *Journal of the American Chemical Society,* **87,** 395 (1965).

Solvolysis of cyclopropyl tosylates does in fact lead to products derived from allyl cation,[7] but the free cyclopropyl cation is not involved as an intermediate. Instead, rearrangement is concerted with ionization, as in many of the solvolyses discussed in Special Topic 3. Under these circumstances disrotatory opening still occurs, but in addition the direction of rotation is such as to move the electrons of the σ bond being broken to a side opposite that of the leaving group.

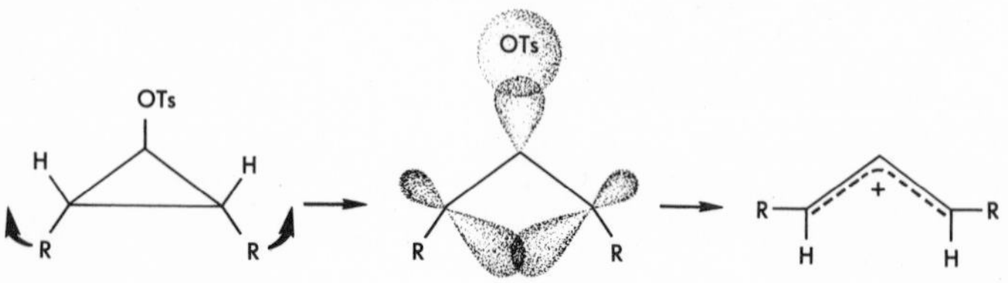

The reason for this preference is related to that involved in the general rule (Chapter 3) that S_N2 reactions go by backside displacement. A particularly striking example[8] is found in the relative acetolysis rates of the two bicyclohexyl tosylates (I) and (II). At 100°, I reacts 2,500,000 times as rapidly as does II.

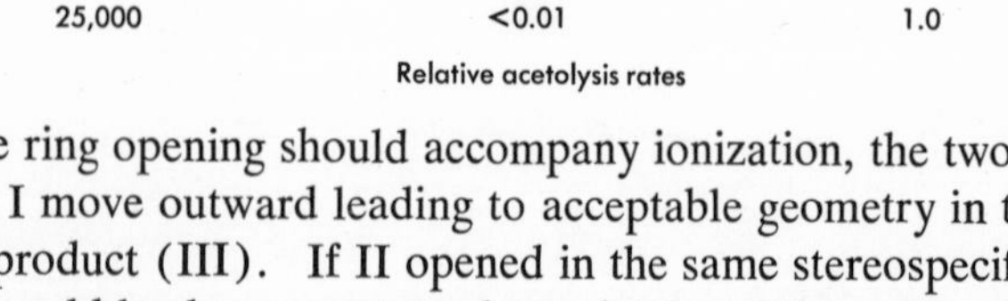

I 25,000 | II <0.01 | 1.0

Relative acetolysis rates

Since ring opening should accompany ionization, the two hydrogens in I move outward leading to acceptable geometry in the allyl cation product (III). If II opened in the same stereospecific fashion, it would lead to an extremely strained cyclohexenyl cation IV.

7. C. H. De Puy, L. G. Schnack, and J. W. Hausser, "The Solvolysis of Cyclopropyl Tosylates," *Journal of the American Chemical Society*, **88**, 3343 (1966).
8. U. Schöllkopf, K. Fellenberger, M. Patsch, P. Schleyer, T. Su, and G. van Dine, "Acetolyse von Endo- und Exo-Bicyclo-(n,1,0)-alkyltosylaten," *Tetrahedron Letters*, 3639 (1967).

I —disrotation / backside displacement→ III

II —disrotation / backside displacement→ IV

Photochemical Electrocyclic Reactions

In Chapter 8 we described the photostationary interconversion of a hexatriene and a cyclohexadiene. Comparison of that case with the thermal reaction (3) just discussed shows that the photochemical process is conrotatory while the thermal reaction is disrotatory. Furthermore, the thermal opening of a cyclobutene to a butadiene in reactions (1) and (2) was conrotatory, while the photochemical closure of the butadiene to a cyclobutene was disrotatory. The difference between the two kinds of electrocyclizations is general.

Let us consider the simple case of the photocyclization of a butadiene to cyclobutene. The photoexcited butadiene (S_1 or T_1) has two electrons in Ψ_1 and one each in Ψ_2 and Ψ_3. As closure occurs this pattern will persist, with one fully-occupied and two singly-occupied orbitals, and the system will begin to resemble an excited state of the product. At some reasonably late stage energy is lost and the ground-state product is formed. An allowed stereoelectronic pathway is one which permits the *excited* starting material to transform smoothly into the *lowest excited state* of the product.

As Figure 8–5 shows, singly-occupied π and π^* of excited cyclobutene can be derived from Ψ_3 and Ψ_2 of excited butadiene (the

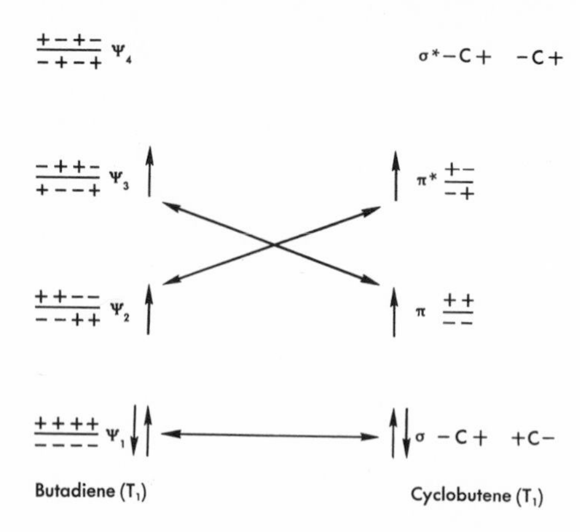

FIGURE 8–5
Correlation of states in a photochemical cyclobutene-butadiene interconversion.

triplet is shown, but the same is of course true for S_1). The σ bond now comes from Ψ_1, and disrotation would be required to get bonding in Ψ_1 between carbons 1 and 4.

The general pattern in photochemical electrocyclic conversions between lowest excited states is that the two singly-occupied orbitals on each side of the reaction correlate (e.g., Ψ_2 and Ψ_3 with π and π^*) and that the σ bond correlates not with the highest occupied orbital but with the highest *fully-occupied* orbital. Thus a diagram similar to Figure 8–5 can be constructed for the hexatriene-cyclohexadiene interconversion. In the lowest excited state of hexatriene, one of the electrons in Φ_3 (cf. the next-to-last section for the hexatriene orbitals) is promoted to Φ_4. The new σ bond must be constructed from Φ_2, the highest fully-occupied orbital, and this requires conrotation as is experimentally observed.

The σ orbital in a photochemical electrocyclic reaction is correlated with the π orbital of the open chain isomer immediately below the one used in a thermal reaction. Thus thermal and photochemical processes must always have different rotatory preferences, since the relative symmetries of the two terminal atoms in a polyene alternate in π orbitals of increasing energy. This theorem follows from the presence of an additional node in each higher π orbital and is easily confirmed by a glance at the hexatriene or butadiene orbitals.

The ideas involved in orbital correlation are of wide generality, and they have been applied to a variety of processes in addition to those discussed here. As with all such unifying concepts they have not only rationalized old observations but also stimulated a flood of new experiments.

INDEX